T0275589

This controversial book examines one of the most important questions of modern cosmology: how much matter is there in the universe? This issue affects theories of the origin and evolution of the universe as well as its geometrical structure and ultimate fate. The authors discuss all the relevant cosmological and astrophysical evidence and come to the conclusion that the balcance of arguments presently lies with a density of around twenty per cent of the critical density required for the Universe to ultimately recollapse. Because the arguments presented constitute a 'state-of-the-art analysis of the observational and theoretical arguments surrounding primordial nucleosynthesis, large-scale structure formation the cosmic microwave background radiation and the existence of case matter, this study provides the reader with an indispensable introduction to the most exciting recent developments in modern cosmology.

Written by two eminent cosmologists, this topical and provocative book will be esential reading for all cosmologists and atrophysicists.

CAMBRIDGE LECTURE NOTES IN PHYSICS 7

General Editors: P. Goddard, J. Yeomans

Is the Universe Open or Closed?

CAMBRIDGE LECTURE NOTES IN PHYSICS

1. Clarke: The Analysis of Space-Time Singularities
2. Dorey: Exact S-Matrices in Two Dimensional Quantum Field Theory
3. Sciama: Modern Cosmology and the Dark Matter Problem
4. Veltman: Diagrammatica – The Path to Feynman Rules
5. Cardy: Scaling and Renormalization in Statistical Physics
6. Heusler: Black Hole Uniqueness Theorems
7. Coles & Ellis: Is the Universe Open or Closed?

Is the Universe Open or Closed?

PETER COLES
Queen Mary & Westfield College, University of London

GEORGE ELLIS
University of Cape Town

PUBLISHED BY THE PRESS SYNDICATE OF THE UNIVERSITY OF CAMBRIDGE
The Pitt Building, Trumpington Street, Cambridge CB2 1RP United Kingdom

CAMBRIDGE UNIVERSITY PRESS
The Edinburgh Building, Cambridge CB2 2RU, United Kingdom
40 West 20th Street, New York, NY 1001–4211, USA
10 Stamford Road, Oakleigh, Melbourne 3166, Australia

First published 1997

A catalogue record for this book is available from the British Library

Library of Congress Cataloguing in Publication data

Coles, Peter.
Is the universe open or closed? / Peter Coles, George Ellis.
p. cm. – (Cambridge lecture notes in physics : 7)
Includes bibliographical references and index.
1. Cosmology. 2. Astrophysics. 3. Large scale structure
(Astronomy) 4. Cosmic background radiation. I. Ellis,
George Francis Rayner. II. Title. III. Series.

QB981.C646 1997
523.1–dc20 96–412.5 CIP

ISBN 0 521 56689 4 (pbk.)

Transferred to digital printing 2003

Contents

Preface

This book had its origins in a coffee-time discussion in the QMW common room in 1993, in the course of which we discussed many issues pertaining to the material contained here. At the time we had this discussion, there was a prevailing view, particularly among cosmologists working on inflationary models, that the issue of the density of matter in the Universe was more-or-less settled in favour of a result very near the critical density required for closure. Neither of us found the arguments made in this direction to be especially convincing, so we determined at that time to compile a dossier of the arguments – theoretical and observational, for and against – in order to come to a more balanced view. This resulted in a somewhat polemical preprint, 'The Case for an Open Universe', which contained many of the arguments we now present at greater length in this book, and which was published in a much abridged form as a review article in *Nature* (Coles & Ellis 1994).

The format of a review article did not, however, permit us to expand on the technical aspects of some of the arguments, nor did it permit us to foray into the philosophical and methodological issues that inevitably arise when one addresses questions such as the origin and ultimate fate of the Universe, and which form an important foundation for the conclusions attained. The need for such a treatment of this question was our primary motivation for writing this book.

Reinforcing this idea was the realisation that the question we sought to address cuts across many different branches of cosmology and that focusing upon this particular issue would allow us to fulfil the useful pedagogical aim of providing a source book of the main current issues in cosmology from a well-defined perspective that takes the observational evidence seriously. We hope that that students as well as established researchers in this field will find the approach interesting and complementary to that taken by more conventional textbooks. This is not, however, intended to be textbook in the sense of being a systematic introduction to all of modern cosmology. Even in this format, space is limited and we have had to quote some standard material without discussion; in such cases we give references to where the source material may be found. It has also proved necessary to quote only selected references to certain topics, particularly large-scale structure and inflationary theory, where the literature is particularly active and crowded. We apologise in advance to those researchers whose work we have passed over.

As it happens, the three years since we first discussed this project together have seen a dramatic shift in the consensus on this issue throughout the cosmology community, with many workers now prepared to accept that the balance of the evidence may not reside with the theoretically favoured critical density case. While our original article may have had something to do with this change of mind, the need for a polemical treatment has now correspondingly diminished. Our aim in this book, therefore, is to give as balanced a view as possible of the many and varied arguments surrounding the issue. Where necessary we will concentrate on the uncertainties and introduce counter-arguments to the accepted wisdom, but we hope to do this fairly and in the interest of giving all the evidence a fair hearing. It is not our aim to give the reader a doctrinaire statement of absolute truth, but to summarise the evidence and, as far as is possible, to indicate where the balance of the argument lies at present. There is still a need for such an analysis.

Acknowledgements

Peter Coles gratefully acknowledges the award of a PPARC Advanced Fellowship during the period in which this book was being written, and thanks PPARC for providing travel and subsistence expenses for a trip to Cape Town in 1995 during which he found solace in writing after England's traumatic defeat by the All Blacks in the Rugby World Cup.

George Ellis thanks the FRD (South Africa) for financial support, and the South African rugby team (the Amabokke) for their World Cup victory, inspired by Nelson Mandela.

Both authors are grateful to a large number of individuals for their advice, encouragement and enjoyable scientific discussions relating to the material in this book. Prominent among these are: Bernard Carr, Richard Ellis, Carlos Frenk, Craig Hogan, Bernard Jones, Andrew Liddle, Jim Lidsey, Jim Peebles, Francesco Lucchin, John Peacock, Manolis Plionis, Martin Rees, Ian Roxburgh and Dennis Sciama.

We are also grateful to Adam Black at CUP for his friendly advice and encouragement and his brave attempts to keep us on target to meet the deadline.

1
Introduction

1.1 The issue

The issue we plan to address in this book, that of the average density of matter in the universe, has been a central question in cosmology since the development of the first mathematical cosmological models. As cosmology has developed into a quantitative science, the importance of this issue has not dimininished and it is still one of the central questions in modern cosmology.

Why is this so? As our discussion unfolds, the reason for this importance should become clear, but we can outline three essential reasons right at the beginning. First, the density of matter in the universe determines the *geometry* of space, through Einstein's equations of general relativity. More specifically, it determines the curvature of the spatial sections†: flat, elliptic or hyperbolic. The geometrical properties of space sections are a fundamental aspect of the structure of the universe, but also have profound implications for the space-time curvature and hence for the interpretation of observations of distant astronomical objects. Second, the amount of matter in the universe determines the rate at which the expansion of the universe is decelerated by the gravitational attraction of its contents, and thus its future state: whether it will expand forever or collapse into a future hot big crunch. Both the present rate of expansion and the effect of deceleration also need to be taken into account when estimating the *age* of the universe. The importance of stellar and galactic ages as potential tests of cosmological theories therefore hinges on reliable estimates on the

† One can think of the spatial sections as being like frames of a movie, their continuing succession building up a space-time (Ellis & Williams 1988). The difference is that these frames are intrinsically three-dimensional (and may be curved!). Cosmology provides us with a natural choice for the time coordinate: cosmological proper time, defined below.

mean matter density. Indeed the very viability of our models is threatened if we cannot attain consistency here. Finally, we would also like to know precisely *what the universe is made of.* There is compelling astrophysical and cosmological evidence for the existence of non-baryonic dark matter, i.e. matter which is not in the form of the material with which we are most familiar: protons and neutrons. Indeed, as we shall see, even a conservative interpretation of the data suggests that the bulk of the matter we can infer from observations may not be made of the same stuff that we are. As well as its importance for cosmology, this issue also has more human aspects, if we accept the implication that the material of which we are made is merely a tiny smattering of contaminant in a sea of quite different dark matter.

Historical attempts to determine this fundamental physical parameter of our universe have been fraught with difficulties. Relevant observations are in any case difficult to make, but the issue is also clouded by theoretical and 'aesthetic' preconceptions. In the early days of modern cosmology, Einstein himself was led astray by the conviction that the universe had to be static (i.e. non-expanding). This led him to infer that the universe must be prevented from expansion or contraction by the presence of a cosmological constant term in the equations of general relativity (Einstein 1917; later he called this a great mistake). The first modern attempt to compute the mean density of the matter component was by de Sitter (1917) whose result was several orders of magnitude higher than present estimates, chiefly because distances to the *nebulae* (as galaxies where then known) were so poorly determined. These measurements were improved upon by Hubble (1926), Oort (1932) and Shapley (1934), who estimated the mean density of matter in the universe to be on the order of $\rho \simeq 10^{-30}$ g cm^{-3}. It is interesting to note that as early as 1932, Oort realised that there was evidence of significant quantities of dark matter, whose contribution to the total density would be very difficult for astronomers to estimate. This point was stressed by Einstein (1955):

> ...one can always give a lower bound for ρ but not an upper bound. This is the case because we can hardly form an opinion on how large a fraction a fraction of ρ is given by astronomically unobservable

(not radiating) masses.

In a classic paper, Gott *et al.* (1974) summed up the empirical evidence available at the time, including that for the dark matter, and concluded that the universe had a sufficiently low density for it to be *open*, i.e. with negatively curved spatial sections and with a prognosis of eternal expansion into the future.

The most notable developments in recent years have been (i) on the observational side, the dynamical detection of dark matter from the behaviour of galactic rotation curves, and (ii) on the theoretical side, the introduction by Guth (1981) of the concept of *cosmic inflation* into mathematical models of the universe. We shall discuss inflation further in Chapter 2, but for now it suffices to remark that it essentially involves a period of extremely rapid expansion in the early universe which results in a 'stretching' of the spatial sections to such an extent that the curvature is effectively zero and the universe is correspondingly flat. If the cosmological constant is zero, this implies the existence of a critical density† of matter: $\rho \simeq 10^{-29}$ g cm^{-3} today on average (see §1.2.4).

But is there sufficient matter in the universe for the flat model to be appropriate? There is a considerable controversy raging on this point and – it has to be said – the responsibility for this lies mainly with theorists rather than observers. Without inflation theory, it is probable that most cosmologists would agree to a value of the mean cosmological density of around 10 to 30 per cent of the critical value, but would have an open mind to the existence of more unseen matter than this figure represents, in line with the comments made by Einstein (1955). It is our intention in this book to explore the evidence on this question in as dispassionate a way as possible. Our starting point is that this issue is one that *must* be settled empirically, so while theoretical arguments may give us important insights as to what to expect, they must not totally dominate the argument. We have to separate theoretical prejudice from observational evidence, and ultimately decide the answer on the basis of data about the nature of the real universe. This is not as simple a task as it may seem, because every relevant observa-

† Although this density of matter is sufficient to close the universe, it is, by our standards, a very good vacuum. It corresponds to on the order of one hydrogen atom per cubic metre of space.

tion depends considerably on theoretical interpretation, and there are many subtleties in this. What we are aiming for, therefore, is not a dogmatic statement of absolute truth but an objective discussion of the evidence, its uncertainties and, hopefully, a reasonable interpretation of where the balance of the probabilities lies at present, together with an indication of the most hopeful lines of investigation for arriving at a firm conclusion in the future.

1.2 Cosmological models

We now have to introduce some of the basics of cosmological theory. This book is not intended to be a textbook on basic cosmology, so our discussion here is brief and focusses only on the most directly relevant components of the theory. For more complete introductions see, for example, Weinberg (1972), Kolb & Turner (1990), Narlikar (1993), Peebles (1993) and Coles & Lucchin (1995).

1.2.1 The nature of cosmological models

In this section we shall introduce some of the concepts underpinning the standard big-bang cosmological models†. Before going into the details, it is worth sketching out what the basic ideas are that underlie these cosmological models. Most importantly, there is the realisation that the force of Nature which is most prominent on the large scales relevant to cosmology is gravity‡. The best theory of gravity that we have is Einstein's general theory of relativity. This theory relates three components:

(i) a description of the space-time geometry;

(ii) equations describing the action of gravity;

(iii) a description of the bulk properties of matter.

We shall discuss these further in the following subsections.

† For alternative models, see Ellis (1984a) and Wainwright & Ellis (1996).
‡ It would be electromagnetism if there were a net charge separation on galactic scales.

1.2.2 Geometry: the cosmological principle

The fundamental principle upon which most cosmological models are based is the *cosmological principle*, which is that the universe, at least on large scales, is homogeneous and isotropic. This assumption makes the description of the geometry of cosmological models much simpler than many other situations in which general relativity is employed. One should admit at the outset, however, that one cannot make a watertight case for the global homogeneity and isotropy of the universe (Ellis *et al.* 1996). We know that the universe is quite inhomogeneous on scales up to at least 100 Mpc. The near-homogeneity of the X-ray background, the counts of distant radio sources, and the cosmic microwave background anisotropy limits (Chapter 7) offer at least some circumstantial evidence that the distribution of material on large scales may be roughly homogeneous on scales much larger than this (e.g. Peebles 1993; Stoeger *et al.* 1995a). More recently, large-scale surveys of galaxies have begun to probe depths where one can see, qualitatively at least, the emergence of a homogeneous pattern (Shectman *et al.* 1996). One hopes that this observed behaviour is consistent with the usual treatment of large-scale structure in terms of small perturbations to a globally homogeneous and isotropic background model. We shall discuss this further in Chapter 6. We will, however, take the uncertainty surrounding the detailed application of the cosmological principle on board and, in Chapter 8, explore some issues pertaining to the role of inhomogeneity in cosmology.

Assuming the cosmological principle holds, we first wish to describe the geometrical properties of space-times compatible with it. It turns out that all homogeneous and isotropic space-times can be described in terms of the Friedman–Robertson–Walker (FRW) line element

$$\mathrm{d}s^2 = c^2\mathrm{d}t^2 - a^2(t)\left(\frac{\mathrm{d}r^2}{1-\kappa r^2} + r^2\mathrm{d}\theta^2 + r^2\sin^2\theta\mathrm{d}\phi^2\right), \qquad (1.1)$$

where κ is the spatial curvature, scaled so as to take the values 0 or ±1, and $u^\alpha = \delta^\alpha_0$ is the average matter four-velocity (defining the world lines of *fundamental observers*). The case $\kappa = 0$ represents flat space sections, and the other two cases are space sections of constant positive or negative curvature, respectively. The time coordinate t is called *cosmological proper time* and it is singled out as a preferred time coordinate by the property of spatial homogeneity: observers can set their clocks according to the local density of matter, which is constant on space-like surfaces orthogonal to the matter four-velocity. The quantity $a(t)$, the *cosmic scale factor*, describes the overall expansion of the universe as a function of time. An important consequence of the expansion is that light from distant sources suffers a cosmological redshift as it travels along a null geodesic in the space-time: $ds = 0$ in equation (1.1). If light emitted at time t_e is received by an observer at t_0 then the redshift z of the source is given by

$$1 + z = \frac{a(t_0)}{a(t_e)}. \tag{1.2}$$

1.2.3 The Friedman equations

The dynamics of an FRW universe are determined by the Einstein gravitational field equations, which can be written, in tensor notation, in the form

$$G^{\nu}_{\mu} = 8\pi G T^{\nu}_{\mu}, \tag{1.3}$$

where T^{ν}_{μ} is the energy-momentum tensor describing the contents of the universe. With this geometry the matter stress-tensor necessarily has the form of a perfect fluid, with $\rho = T^0_0$, $\mathrm{p} = -\frac{1}{3}T^{\alpha}_{\alpha}$, because the matter four-velocity is $u^{\alpha} = \delta^{\alpha}_0$. The Einstein equations then simplify to

$$3\left(\frac{\dot{a}}{a}\right)^2 = 8\pi G\rho - \frac{3\kappa c^2}{a^2} + \Lambda, \tag{1.4}$$

$$\frac{\ddot{a}}{a} = -\frac{4\pi G}{3}\left(\rho + 3\frac{p}{c^2}\right) + \frac{\Lambda}{3}, \tag{1.5}$$

$$\dot{\rho} = -3\frac{\dot{a}}{a}\left(\rho + \frac{p}{c^2}\right). \tag{1.6}$$

These equations determine the time evolution of the cosmic scale factor $a(t)$ (the dots denote derivatives with respect to cosmological proper time t) and therefore describe the global expansion or contraction of the universe. The first equation (the Friedman equation) is a first integral of the other two. In the early phases of the big-bang, the universe is dominated by radiation or relativistic particles for which $p = \rho c^2/3$, while for late times (including now) it is matter-dominated, so that $p \simeq 0$. The crossover between these two regimes occurs when the scale factor a is between 10^{-3} and 10^{-5} of its present value, depending on the density of matter.

In inflationary models of the early universe, there exists a short period of time in which the dynamics of the universe are determined by the action of a scalar field which has an effective equation of state of the form $p = -\rho c^2$ if it is in the slow-rolling regime. We shall discuss this option further in Chapter 2. The cosmological constant Λ, which many cosmologists believe to be exactly zero, changes the acceleration of the universe compared to models containing only matter and/or radiation. A scalar field with $p = -\rho c^2$ behaves in essentially the same way as a cosmological constant term.

It is a property of the homogeneous and isotropic expansion of the universe around every point that these models can easily reproduce Hubble's law for the recession of galaxies:

$$v = H_0 r, \tag{1.7}$$

where r is the proper distance of a galaxy and v is its apparent recession velocity, inferred from the redshifting of spectral lines. The parameter H_0 is called the Hubble constant. In terms of the scale factor, it is straightforward to see that

$$H_0 = (\dot{a}/a)_{t=t_0}, \tag{1.8}$$

with the suffix referring to the present time t_0; in general the expansion rate $H(t) = \dot{a}/a$. The actual value of H_0 is not known with any great accuracy, but is probably in the range 40 km s^{-1} Mpc^{-1} $< H_0 <$ 90 km s^{-1} Mpc^{-1}. This uncertainty is usually parametrised by writing $H_0 = 100h$ km s^{-1} Mpc^{-1}, where h is dimensionless. We shall discuss the observational evidence pertaining to H_0 in Chapter 3: recent estimates suggest $h \simeq 0.7$ (e.g. Freedman *et al.* 1994; Pierce *et al.* 1994) but, to be safe, conservative limits on h are $0.4 \leq h \leq 0.9$.

1.2.4 Open, closed and flat cosmologies

We are now in a position to introduce the relationship between the density of matter and the curvature of the spatial sections we discussed in §1.1. The important parameter for determining the long-term evolution of an FRW universe is the *density parameter*, Ω, which is defined to be

$$\Omega = \frac{8\pi G\rho}{3H^2} = \frac{\rho}{\rho_{\rm cr}}, \tag{1.9}$$

in other words, the ratio of the actual density of the universe to a critical value $\rho_{\rm cr}$. The present value of this critical density depends on H_0:

$$\rho_{\rm cr} = \frac{3H_0^2}{8\pi G} \simeq 1.9 \times 10^{-29} h^2 \,{\rm gm\,cm^{-3}}. \tag{1.10}$$

The value (1.10) is the yardstick against which we shall measure the various determinations of ρ_0 and we shall henceforth give most of our estimates in terms of Ω-values.

If $\Omega > 1$ then $\kappa = +1$ – elliptic spatial sections – and the universe will recollapse to a second singularity (a 'big crunch'); if $\Omega < 1$ then $\kappa = -1$ – hyperbolic spatial sections – and it will expand forever with an ever-decreasing density. In between, $\Omega = 1$ corresponds to the flat $\kappa = 0$ universe favoured by some inflationary models for the early universe (e.g. Guth 1981; Linde 1982). The relationship between κ and Ω becomes more complicated if we allow the cosmological constant to be non-zero.

1.2.5 The equation of state

The Friedman equation (1.4) becomes determinate when we select an equation of state of the form $p = p(\rho)$ for the matter. Particularly relevant examples are $p = 0$ for pressureless matter (or 'dust') and $p = \frac{1}{3}\rho c^2$ for relativistic particles or radiation. One can then use the conservation equation (1.6) to determine $\rho(a)$. The result for pressureless matter is $\rho = M/a^3$, while for radiation it is $\rho = M/a^4$, where M is constant.

Alternatively, one can represent the matter in terms of a scalar field whose evolution is governed by a potential $V(\phi)$ – this is the case in inflationary models in particular (see §2.3 below). In such a case the effective density and pressure are given by

$$\begin{aligned} \rho c^2 &= \frac{1}{2}\dot{\phi}^2 + V(\phi), \\ p &= \frac{1}{2}\dot{\phi}^2 - V(\phi). \end{aligned} \tag{1.11}$$

We discuss some aspects of scalar field cosmologies in Chapter 2.

If $\Omega = 1$ the scale factor evolves according to: (i) $a(t) \propto t^{2/3}$ for pressureless matter, and (ii) $a(t) \propto t^{1/2}$ for radiation. It is plausible that the universe is radiation-dominated from the time of nucleosynthesis to about decoupling, and matter-dominated from then on, and so early on behaves like a radiation universe and later behaves like a matter universe. In a universe with $\Omega < 1$, the universe expands roughly like this until it reaches a fraction Ω of its present size and then goes into free expansion with $a(t) \propto t$. If $\Omega > 1$, it expands more slowly than the critical density case, and then recollapses. This behaviour has implications for structure formation, as we shall see in Chapter 6.

1.2.6 *Useful formulae*

It is often convenient to express the dynamical equations in terms of H and Ω rather than t and ρ. From the Friedman equation, referred to the reference time t_0, one can write

$$\left(\frac{\dot{a}}{a_0}\right)^2 - \frac{8\pi G\rho}{3}\left(\frac{a}{a_0}\right)^2 = H_0^2(1-\Omega_0) = -\frac{\kappa c^2}{a_0^2} = \kappa_0. \qquad (1.12)$$

One can also write

$$\left(\Omega^{-1} - 1\right)\rho(t)a(t)^2 = -\kappa c^2 = \left(\Omega_0^{-1} - 1\right)\rho_0(t_0)a(t_0)^2, \qquad (1.13)$$

which we shall find useful later, in Chapter 2. The appropriate relation between curvature and Ω_0 in the presence of the Λ-term is

$$\frac{\kappa c^2}{a_0^2} - \frac{\Lambda c^2}{3} = H_0^2(\Omega_0 - 1) \qquad (1.14)$$

(this is just the Friedman equation again, evaluated at $t = t_0$).

1.2.7 *The big-bang model*

Most cosmologists accept the big-bang model as basically correct: the evidence in favour of it is circumstantial but extremely convincing (Peebles *et al.* 1991). In particular, we can quote the agreement of predicted and observed abundances of atomic light nuclei (Chapter 4) and the existence of the microwave background radiation (Chapter 7), a relic of the primordial fireball. It is important to remember, however, that the big-bang model is not a complete theory. For example, it does not specify exactly what the matter contents of the universe are, nor does it make a prediction for the expansion rate H_0. It also breaks down at the singularity at $t = 0$, where the equations become indeterminate: 'initial conditions' therefore have to be added at some fiducial early time where the equations are valid. Theorists therefore have considerable freedom to play around with parameters of the model in order to fit observations. But the big-bang model does at least provide a conceptual framework within which data can be interpreted and analysed, and allows meaningful scientific investigation of the thermal history of the universe from times as early as the Planck epoch $t = t_{\mathrm{P}}$, where

$$t_{\mathrm{P}} = \left(\frac{\hbar G}{c^5}\right)^{1/2} \simeq 10^{-43}\,\mathrm{s} \tag{1.15}$$

(e.g. Kolb & Turner 1990; Linde 1990).

1.3 Cosmological criteria

It will be apparent to anyone who has glanced at the literature, including the newspapers, that there is a great deal of controversy surrounding the issue of Ω_0, sometimes reinforced by a considerable level of dogmatism in opposing camps. In understanding why this is the case, it is perhaps helpful to note that much of the problem stems from philosophical disagreements about which are the appropriate criteria for choosing an acceptable theory of cosmology. Different approaches to cosmology develop theories aimed at satisfying different criteria, and preferences for the different approaches to a large extent reflect these different initial goals. It would help to clarify this situation if one could make explicit the issues relating to choices of this kind, and separate them from the more 'physical' issues that concern the interpretation of data. Pursuing this line of thought, we now embark on a brief philosophical diversion which we hope will initiate a debate within the cosmological community†.

† Some cosmologists in effect claim that there is no philosophical content in their work and that philosophy is an irrelevant and unnecessary distraction

To provide a starting point for this debate we suggest that the following criteria, derived from criteria that are used in a wider context for scientific theories in general, encapsulate the essentials of this issue: theories should be chosen for:

1. **Satisfactory structure.** This comprises such notions as

   (a) *internal consistency*;
   (b) *simplicity* (in the sense of Ockham's razor: one should choose the minimum number of explanatory elements consistent with the observations);
   (c) *aesthetic appeal* (or, equivalently, *beauty* or *elegance* of a theory).

2. **Intrinsic explanatory power.** Important aspects of this criterion are:

   (a) *logical tightness* (i.e. freedom from arbitrariness of functional relationships or parameters in the theory or model);
   (b) *scope*, including the ability to unify (or at least relate) otherwise separate themes, and also the breadth of implications of the theory;
   (c) *probability* of the theory or model with respect to some well-defined measure.

3. **Extrinsic explanatory power, or relatedness.** This incorporates such concepts as:

   (a) *connectedness* (i.e. the extent of the relationship of the theory to the rest of physics);
   (b) *extendibility* (i.e. the ability of the theory to provide a basis for further development).

4. **Observational and experimental support.** This is to be interpreted in two senses:

   (a) *testability*, the ability of the theory to make predictions that can be tested;
   (b) *confirmation*, the extent to which the theory has been supported by tests that have been made.

from their work as scientists. We claim that they are, whether they like it or not, making philosophical (and, in many cases, metaphysical) assumptions, and it is better to have these out in the open than hidden.

One can imagine a kind of rating system which judges cosmological models against each of these criteria. The point is that cosmologists from different backgrounds implicitly assign a different weighting to each of them, and therefore end up trying to achieve different goals to others. There is a possibility of both positive and negative ratings in each of these areas.

Note that we have rejected categories such as 'importance', 'intrinsic interest' and 'plausibility' – insofar as they have any meaning apart from personal prejudice, they should be reflected in the categories above, and could perhaps be defined as aggregate estimates following on from the proposed categories. Category 1(c) ('beauty') is difficult to define objectively but nevertheless is quite widely used, and seems independent of the others; it is the one that is most problematic† in terms of getting agreement, so some might wish to omit it. One might think that category 1(a) ('logical consistency') would be mandatory, but this is not so, basically because we do not yet have a consistent 'Theory of Everything'. Again one might think that negative scores in 4(b) ('confirmation') would disqualify a theory but, again, that is not necessarily so, because measurement processes, statistics and results are all to some extent uncertain and can all therefore be queried.

The idea is that even when there is disagreement about the relative merits of different models or theories, there is a possibility of agreement on the degree to which the different approaches could and do meet these various criteria. Thus one can explore the degree to which each of these criteria is met by a particular cosmological model or approach to cosmology. We suggest that one can distinguish five broadly different approaches to cosmology, roughly corresponding to major developments at different historical epochs (Ellis 1990, 1993):

(A) **geometrical**, corresponding to the idea of *cosmography* – one aims to specify the distribution of matter and space-time geometry

† An interesting insight into this kind of argument was given by Martin Rees in the 1996 National Astronomy Meeting in Liverpool. He compared and contrasted the apparently 'beautiful' circular orbit model of the Solar System with the apparently 'ugly' elliptic orbits found by Kepler. Only after Newton introduced his theory of gravitation did it become clear that beauty in this situation resided in the inverse-square law itself, rather than in the outcomes of that law.

without use of a dynamical theory (e.g. Hubble 1934, Milne 1936);
(B) **observational**, concentrating on determination of space-time geometry by observation of discrete sources, and taking the field equations into account (e.g. Sandage 1961, 1970);
(C) **astrophysical** (post-decoupling and decoupling), including observations and analyses of background radiation and intergalactic matter;
(D) **astroparticle**, split into D1, from decoupling back to nucleosynthesis, and D2, prior to nucleosynthesis but after the Planck time;
(E) **quantum** (prior to the Planck time), including attempts to specify a law of initial conditions for universe models.

These approaches† are not completely independent of each other, but any particular model will tend to focus more on one or other aspect and may even completely leave out others. Comparing them with the criteria above, one ends up with a 'star rating' system something like that shown in Table 1.1, in which we have obviously applied a fairly arbitrary scale to the assignment of the ratings!

This book will take into account all these criteria, and try to make explicit which of them come into play at the various stages of our analysis. However our underlying essential view is that without a solid basis of experimental support [4(b)], or at least the possibility of confirmation [4(a)], a proposed theory is not on solid ground; one can say what one likes and cannot be proved wrong, and so is free from the normal constraints of scientific discipline. In this sense our approach contrasts with a major thrust in modern astrophysical thinking which emphasises criteria (2) and (3) at the expense of (4). This will become clear as we proceed with our analysis.

1.4 Preliminary discussion of estimates

After the preceding philosophical detour, let us now resume the thread of argument leading to the discussion of Ω we present in

† We might include a further category relating to the existence of life and the anthropic principle. We refrain from doing so, although we regard these issues as important, in order to avoid controversy diverting the reader from the main lines of argument of this book. We may return to this theme in another book.

Table 1.1. *Four-star ratings for approaches to cosmology.*

	Approach	A	B	C	D1	D2	E
	Criterion						
1.	Structure						
(a)	consistency	****	***	**	***	**	*
(b)	simplicity	****	*	*	*	*	*
(c)	aesthetic appeal	****	*	**	**	**	**
2.	Explanatory power						
(a)	tightness	***	**	**	***	***	*
(b)	scope	*	**	***	***	****	****
(c)	probability	*	*	*	****	***	
3.	Relatedness						
(a)	connectedness	*	**	****	****	****	****
(b)	extendibility	***	*	****	****	****	****
4.	experimental support						
(a)	testability	***	****	***	***	**	*
(b)	confirmation	**	***	**	***	*	

more detail in later chapters.

The critical density case, which just succeeds in expanding forever, corresponds to the Einstein–de Sitter universe models with flat spatial sections, and occurs when $\Omega_0 = 1$. In recent years, largely as a result of the inflationary universe models, there has been a large body of opinion claiming that the density is very close indeed to that in a critical density universe: $\Omega_0 = 1$ to very high accuracy today, despite the fact that until very recently all the unambiguous astronomical density determinations have suggested a much lower figure.

It has been known for a long time that the directly observed luminous material in the central regions of galaxies contributes only a very small fraction of the critical density: $\Omega_0 \simeq 0.005$. Dynamical estimates of the masses of galaxies using, for example, rotation curves of spiral galaxies, suggest that such galaxies contain large quantities of dark matter which might take the cosmological density up to $\Omega_0 \simeq 0.02$ or so. Estimates of Ω_0 based on the mass-to-light ratios of rich clusters of galaxies and galaxy motions on larger scales give results in the range $0.1 < \Omega_0 < 0.3$, considerably higher than the density inferred from luminous matter. (We shall discuss this latter evidence, as well as more indirect constraints on Ω_0, in much more detail later.) This already leads to the famous dark matter problem: the search for the nature of the non-luminous matter that must be there, according to this dynamical evidence. The problem is enhanced if indeed we live in a critical density universe, for then at least three times as much dark matter must be present.

A solution to this problem may be furnished by the early stages of the evolution of the big bang, where the ambient temperature is sufficiently large to create exotic particles which may survive today as relics. There are sufficient uncertainties in high energy particle physics for there to be many candidate particles, from relatively well-known ones such as neutrinos with small but non-zero mass, to more exotic alternatives produced in supersymmetric theories (e.g. photinos). We discuss these possibilities further in the next chapter. While laboratory searches for dark matter candidates have so far been in vain, in the past few years this theoretical predilection for a high-density universe has been reinforced by various astronomical observations that have been taken as supporting the critical value. Nevertheless the weight of evidence has actually been rather for lower densities (Peebles 1986; Coles & Ellis 1994; Krauss & Turner 1995; Ostriker & Steinhardt 1995; Ellis *et al.* 1996). Some enthusiasts, however, have seemed not to be very concerned about the evidence, even going so far as to claim the density *must* take the critical value, and that alternatives are to be regarded as 'cranky' as the following extract from Dennis Overbye's *Lonely Hearts of the Cosmos* (1993), referring to remarks made by a prominent Chicago physicist about the value of Ω_0, demonstrates:

> If you understood anything at all about grand unified theories and inflation, you would realize that omega had to be 1.0. A true physicist could think no other way; it was the paradigm. The cosmologist's job was to reconcile the observations with that number.
>
> We were outside his office talking to one of his graduate students, a young woman who was doing numerical simulations of the

universe. She was doing great, he said, but why was she doing these runs with omega equal to 0.2?

That was what Davis and White did, she answered.

Schramm told her to get rid of those runs. 'You're thinking like an astronomer instead of like a physicist,' he snorted, adding, 'Simon White never understood inflation.'

As we have set out above, we do not accept such a domination of theoretical predeliction over observational test.

1.5 Plan of the argument

In our view, the question of Ω is far from resolved. Indeed, as we go through the evidence it will become clear that many of the relevant arguments are strongly contradictory. Given the importance of this issue, in terms of both the viability of the current generation of cosmological models and the general principles involved in testing cosmological models, and the recent dramatic increase in the number of relevant observations, we believe it is an appropriate time to review the evidence, and to see if the case for a critical density universe can be made beyond reasonable doubt. In doing this, we shall adopt as model-independent an approach as is possible. Where an argument is model-dependent (as, indeed, all arguments are to some extent), we will try to indicate which of the model assumptions might be the most fragile. We shall also suggest possible sources of error in the various observational estimates and try to quantify the possible magnitude of these errors. In some cases, we will assign errors which are much larger than those quoted by the scientists who published the estimates since we feel that there has been a tendency to exaggerate the reliability of some of the results quoted, particularly those claiming a high density†.

Inter alia the purpose of doing this is to emphasise that this is indeed an experimental question, where theory – no matter how dear it may be to us – will eventually have to bow to the experimental evidence. It may be that theoretical prejudices in favour of the high density models will one day be confirmed; if so

† This *must* be true in the case of the other cosmological parameter, H_0, where error bars on different estimates are frequently mutually exclusive.

that will be a great triumph for those theories that imply $\Omega \simeq 1$. Our conclusion, however, will be that at the present time we have not attained that situation: there is plenty of room for reasonable doubt, and low-density models are at least as good as, and probably better than, the most successful critical density ones at explaining the whole range of relevant observations.

The outline of the argument is as follows. First, in Chapter 2, we shall address the theoretical issues surrounding matters such as inflation and particle physics arguments. We then go onto a number of more-or-less direct observational arguments. In Chapter 3 we discuss 'classical' cosmological issues: expansion rate, ages, and attempts to measure the deceleration parameter and curvature effects using extragalactic objects as standard candles. Nucleosynthesis – and the constraints in places on the amount of baryonic material – is discussed in Chapter 4, alongside a review of recent determinations of the relevant light element abundances. In Chapter 5 we look at evidence from a variety of classes of astrophysical argument: this is the evidence for dark matter within observable objects such as galaxies, clusters of galaxies and the intergalactic medium. We then turn, in Chapter 6, to the evidence, perhaps the most circumstantial of all, for dark matter on very large scales as inferred from the statistics of large-scale galaxy clustering and peculiar motions. Chapter 7 contains an account of the use of microwave background anisotropy measurements to infer a value for Ω_0 and the other cosmological parameters: this may turn out to be the route to a precise determination of Ω_0 within the next ten years or so. As we have already mentioned, the issue of inhomogeneity relates to the very foundations of the Friedmann cosmological models, so we end our analysis with Chapter 8, raising some issues connected with the validity (or otherwise) of the cosmological principle. Chapter 9 contains a 'summing up' of the various bits of evidence, in which we adopt a forensic approach, pointing out the uncertainties in and possible objections to the various arguments we have presented. We shall also give our own view of the best prospects for resolving this issue in the relatively near future.

2
Theoretical arguments

As we mentioned in Chapter 1, the main reasons for a predisposition towards a critical density universe are theoretical. We will address these issues carefully, but please be aware at the outset of our view that, ultimately, the question of Ω_0 is an observational question and our theoretical prejudices must bow to empirical evidence.

2.1 Simplicity

In the period from the 1930s to the 1970s, there was a tendency to prefer the Einstein–de Sitter (critical density) model simply because – consequent on its vanishing spatial curvature – it is the simplest expanding universe model, with the simplest theoretical relationships applying in it. It is thus the easiest to use in studying the nature of cosmological evolution. It is known that, on the cosmological scale, spatial curvature is hard to detect (indeed we do not even know its sign), so the real value must be relatively close to zero. Moreover, many important properties of the universe are, to a good approximation, independent of the value of Ω. The pragmatic astrophysicist thus uses the simplest (critical density) model as the basis of his or her calculations – the results are good enough for many purposes (e.g. Rees 1995).

There are, in addition to this argument from simplicity, a number of deeper theoretical issues concerning the Friedman models which have led many cosmologists to adopt a stronger theoretical prejudice towards the Einstein–de Sitter cosmology than is motivated by pragmatism alone.

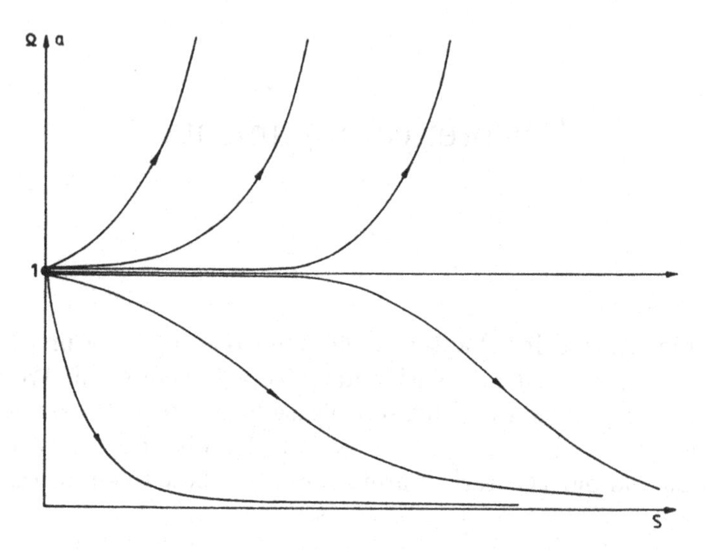

Fig. 2.1 The phase planes for Friedman models with $\gamma > 2/3$. From Madsen & Ellis (1988).

2.2 The flatness problem

The so-called *cosmological flatness problem*, which we describe in some detail below, has cropped up a number of times in different guises during the history of modern cosmology; see, for example, Dicke & Peebles (1979). In the more modern context, it is usually accepted as one of the major motivations for the existence of an epoch of inflation, and an alleged consequence of inflation is that Ω_0 should be very close to unity. Since this issue is clouded by many misconceptions, we shall address it thoroughly here.

2.2.1 Evolution of Ω

The evolution equation (1.5) can be written in the form of an equation for $d\Omega/da$ (e.g. Madsen & Ellis 1988). From this latter equation, phase planes showing the evolution of Ω for different initial conditions can be constructed. An example of such an evolution for a universe with $\gamma > 2/3$ is shown in Figure 2.1; this example serves to illustrate how Ω evolves in a universe dominated by matter or radiation, and therefore shows what happens in a non-inflationary cosmology.

The value of phase planes of this kind is that they can be used to understand the dynamical properties of the equations without resorting to a full solution of the equations. It is immediately obvious from this figure, as indeed it is from the Friedman equations themselves particularly in the form (1.14), that Ω is not a fixed parameter in models of this type (which includes matter-dominated and radiation-dominated examples): it evolves strongly with cosmic time t in such a way that $\Omega = 1$ is an *unstable fixed point.*

2.2.2 Statement of the problem

The phase-plane we have presented means that, if a universe begins with Ω *exactly* equal to unity at the big bang, then $\Omega = 1$ forever. (Of course, these models are all classical and are therefore expected to break down at the Planck time: we should therefore take 'initial conditions' to be given at $t = t_P$ defined by equation (1.16), rather than $t = 0$, but, for the moment, this is not an important point.) Any slight perturbation above or below the value unity, however, leads to a universe in which Ω respectively increases or decreases rapidly with time. Consequently, to get a value of Ω anywhere near unity at the present time (even a factor of a few either way) requires a value extremely close to unity at very early times. To give an example, suppose we take a fiducial early time to be the Planck time, t_P; to end up with Ω_0 of order unity requires $\Omega(t_P) = 1 \pm 10^{-60}$.

The *cosmological flatness problem*, discussed in much more detail in Coles & Lucchin (1995, Chapter 7), arises from the judgement that, within the framework furnished by the standard Friedman models, this 'fine-tuning' is somehow unlikely. The argument, loosely speaking, is that if one imagines that Ω could take any value initially (or perhaps at the Planck time) then the set of choices of this initial condition that lead to $\Omega \simeq 1$ at the present time is, in some sense, of zero measure with respect to the set of all possible choices.

To put this argument another way around, consider the radius of curvature of the universe, which is determined by Ω_0. If $\Omega_0 = 1$ then this radius is infinite. If Ω_0 is not equal to unity exactly, but is some value not too far from unity at the present epoch, then the present curvature radius is larger than the Hubble radius, c/H_0, by some (large) factor. This requires the existence of a large dimensionless ratio, the typical signature of a fine-tuning or coincidence. Since a value exactly equal to unity seems to be unlikely *a priori* according to the argument given above, there seems to be no reason why any particular fine-tuning must hold. Such models are therefore asserted to be unsatisfactory from the point of view of describing our universe.

As we shall see below, where we discuss inflation, this question is usually 'resolved' by appealing to some transient mechanism which can make Ω evolve towards unity for some time, rather than away from it, thus enlarging the set of initial conditions compatible with the present limits on Ω_0 or, alternatively, vastly stretching the curvature scale so that it effectively takes the infinite value associated with the Einstein–de Sitter model.

Notice before we go any further that *all* non-inflationary Friedman models of the type discussed here have Ω arbitrarily close to unity at arbitarily early times. It would appear on these grounds that all possible values of Ω_0 are therefore equally infinitely unlikely, because they all evolve from an infinitely small range of initial values of this parameter. In particular, the 'flatness problem' in this form does not just apply to universes with $\Omega_0 \simeq 1$ but also to those with $\Omega_0 = 0.1$ or $\Omega_0 = 10$: because $\Omega \to 1$ for all these models, any value of Ω_0 requires some amount of fine-tuning if we set initial data at $t = t_P$. It is a subtle issue as to how much is acceptable. However, we can also argue that there is nothing surprising about this at all, as all models start at $\Omega = 1$ in the limit $a \to 0$: such 'fine tuning' is required by the field equations, and can only fail to hold if they fail to hold.

2.2.3 An extended Copernican principle?

The models with $\Omega_0 \neq 1$ have been asserted to violate the most general form of the *Copernican principle*, that we do not occupy a special place in the universe, the argument here being applied to temporal, rather than spatial position. This argument is best stated by considering the curvature radius again. If, say, $\Omega_0 = 0.2$ then the curvature radius exceeds the Hubble radius c/H_0 by a factor which is not particularly large. As time evolves, Ω

will become smaller and smaller and the curvature radius will therefore decrease. The Hubble radius, however, increases with time in these models. If $\Omega_0 = 0.2$ we therefore appear to live at an epoch when when the universe is just becoming curvature-dominated. This argument then asserts that we should prefer an exactly flat universe, because in that case the curvature radius is always infinitely large and there is no selection of a 'special place' for us during the evolution of the universe.

There are three important objections to this line of argument. Firstly, one should always bear in mind that we do to some extent occupy a special temporal position, even in a flat universe, if only because of the requirement that there has to be enough time between the big bang and the present epoch for life to evolve. This observation is an example of the use of the 'weak anthropic principle' (Barrow & Tipler 1986); though often dismissed by cosmologists, this kind of argument is actually essential to the correct formulation of probabilistic arguments in cosmology (Garrett & Coles 1993). The only models in which the Copernican principle, in the sense used here, is exactly true are the steady-state cosmologies (Bondi 1960).

Secondly, and this objection is of general relevance in this context, there is the problem of interpreting the 'coincidences' we have been talking about. Arguments of this type are probability-based and therefore rely explicitly on an appropriate choice of measure. (Coincidences are events which seem 'improbable', given what we know, or at least what we think we know, about the circumstances in which they arise.) In particular, it ought to be clear that there is simply no justification for choosing the initial conditions for Ω arbitrarily according to a uniform measure on the set of positive real numbers. We discuss some pertinent ideas concerning how to assign measures in cosmological models in the next section.

Thirdly, in a $\kappa < 0$ universe, as we shall see in the next section, the most probable value for Ω if we observe 'at a random time' is very close to zero, while if $\kappa > 0$ all decades of Ω greater than 1 are more or less equally probable. Only in the exceptional case $\kappa = 0$ (when $\Omega = 1$ exactly at all times) does the probabilty remain near unity. This case is self-similar, so its evolution is independent of time in a weak sense; but it is also unstable to perturbations in the density, and so is an unviable universe model, as is discussed below. Thus, even a weak extended Copernican principle, based on the time behaviour of dust or radiation models, does not really predict that Ω should be close to unity.

2.2.4 Measures and probabilities

Given the importance of the question of measures in making inferences in cosmology, it is a matter of some regret that this subject has been paid relatively little attention by cosmologists working on the early universe. It is fair to say that there is no general consensus on how this is to be done, despite substantial effort (Gibbons, Hawking & Stewart 1987; Hawking & Page 1988; Ellis 1991; Cho & Kantowski 1994; Coule 1995) and that we are consequently not in a position to make meaningful *a priori* probabilistic arguments about the value of Ω. This point is particularly pertinent when we run into quantum effects near the Planck time, with all the difficulties involved in interpreting the wavefunction of the universe in a meaningul probabilistic way. For the moment, however, we shall restrict ourselves to the interpretation of probabilistic statements in purely classical universe models, i.e. those described simply by the Friedman equations.

The approach to probability which we believe will be most fertile in the future will be that based on the objective Bayesian† interpretation of probability which, we believe, is the only way to formulate this type of reasoning in a fully self-consistent way. In this interpretation, probability represents a generalisation of the notions of 'true' and 'false' to intermediate cases where there is insufficient information to decide with logical certainty between these two alternatives (Cox 1946). Unlike the opposing 'frequentist' view, the Bayesian lends itself naturally to the interpretation of unique events, of which the big bang is an obvious example (Garrett & Coles 1993).

In the Bayesian approach, probability assignments depend explicitly on the acceptance of prior information in the form of a model, in this case a cosmological model. One needs to assign a 'prior' probability that represents this information when making

† For a nice introduction to the benefits of being a Bayesian, see Cousins (1995).

an inference based on observations. Although the means for doing this is still controversial even amongst Bayesians, the most general objective algorithm available appears to be the so-called 'Jaynes principle' (Jaynes 1969). According to this principle, one looks for a measure on the parameter space of the system that possesses the property of invariance under the group of transformations which leave unchanged the mathematical form of the physical laws describing the system. In the absence of any other constraints, the principle of maximum information entropy (a principle of least prejudice) yields a prior probability simply proportional to this measure.

Taking the appropriate model to be a matter-dominated Friedman model as an example, and marginalising over the choice of the related parameter H, Evrard & Coles (1995) have shown that Jaynes' principle leads to a measure for Ω of the form

$$\mu(\Omega) \propto \frac{1}{\Omega|\Omega - 1|}. \tag{2.1}$$

Note that this measure leads to an *improper* (i.e. non-normalisable) prior probability. This can be rectified by bringing in additional information, such as the ages of cosmic objects which rule out high values of both Ω and H. Anthropic selection effects can also be brought to bear on this question: life could never have evolved in a universe with high values of Ω_0 and H_0.

The measure in Ω diverges at $\Omega = 0$ and $\Omega = 1$, the former corresponding to an empty universe without deceleration and the latter to the critical-density Einstein–de Sitter model. These singularities could have been anticipated because these are two fixed points in the evolution of Ω. A model with $\Omega = 1$ exactly remains in that state forever. Models with $\Omega < 1$ evolve to a state of free expansion with $\Omega = q = 0$. Since states with $0 < \Omega < 1$ are transitory, it is reasonable, in the absence of any other information, to infer that the system should be in one of the two fixed states†. (All values of $\Omega > 1$ are transitory.) In the context of this model, therefore, a certain amount of prejudice against intermediate values of

† A useful analogy here is that of a tightrope walker. If one is told that an individual set out to walk along a rope some time in the indefinite past, it is reasonable to infer that, if observed at 'a random time', the walker should be either on the rope or on the ground: it would be very surprising to encounter him/her in mid-fall.

Ω_0 is well motivated by this 'minimal information' approach. To put it another way, the assumption of a constant prior for Ω is not consistent with the assumption of minimal information and therefore represents a *statement of prejudice.* This prejudice may be motivated to some extent by quantum-gravitational considerations that render the classical model inappropriate. Indeed we know that classical cosmological models are expected to break down near the Planck time, so some alternative measure should be specified, and one certainly cannot use this argument to claim that $\Omega \neq 1$ is infinitely improbable.

The crucial point is that unless the model adopted and its associated information are stated explicitly one has no right to assign a prior measure and therefore no right to make any inferences. The measure (2.1) also demonstrates how dangerous it is to talk about Ω_0 'near' unity. In terms of our least-informative measure, values of Ω not exactly equal to unity are actually infinitely far from this value. Furthermore, the probability associated with smaller and smaller intervals of Ω (around unity) at earlier and earlier times need not become arbitrarily small because of the singularity at $\Omega = 1$. Indeed, this measure is constructed in precisely such a way that the probability associated with a given range of Ω_0 is preserved as the system evolves. We should not therefore be surprised to find $\Omega_0 \simeq 1$ at the present epoch even in the absence of inflation, so we do not need to invoke an additional ad hoc mechanism to 'explain' this value. In this sense, *there is no flatness problem* in a purely classical cosmological model.

The argument we have presented here does, however, provide a stronger and more objective foundation for the vague formulation of the Copernican principle we presented in Section 2.2.3. We have addressed the first objection raised there by adopting a coherent formulation of the concept of probability. It should appear unlikely to us that we should live at an epoch when Ω_0 is not very close to zero, not very close to unity and not infinitely large. Since the first and last of these options are ruled out on anthropic grounds†, we appear to be motivated towards Ω_0 being very close to unity. We have not escaped from the third objection, however, because

† Anthropic considerations are not minimally informative in the sense we have used this concept.

the instability of this model is still pertinent if we believe, as we perhaps should, that the universe could have been inhomogeneous at early times.

We believe that this formulation of the problem is a useful example of how such arguments can (and must) be clarified by using a consistent interpretation of the role of probability. On the other hand, the motivation it suggests for a model with Ω_0 arbitrarily close to unity cannot be taken as overwhelmingly compelling because we know that the Friedman equations, along with the measure implied by them, are expected to break down perhaps at the Planck time, and certainly at the big-bang singularity.

2.3 Inflationary models

The inflationary universe scenario (e.g. Guth 1981; Kolb & Turner 1990; Linde 1990; Olive 1990) envisages a dramatic period of exponential inflation through many e-foldings of the scale factor in the very early universe, driven by a scalar field or effective scalar field potential. This has provided a new and much stronger foundation for believing $\Omega_0 \simeq 1$, irrespective of whether or not we believe there is a flatness problem.

2.3.1 The effect of inflation

The basic idea of inflation is that, for some (perhaps short) interval in the early universe, the dynamics of the expansion were dominated by the effect of a scalar field, which behaves as a fluid with an effective equation of state described by $\gamma < 2/3$. A 'realistic' inflationary model begins as a radiation-dominated model, then undergoes a short period of rapid expansion during which $\gamma < 2/3$, then emerges back into a radiative epoch. Subsequently, after further expansion and cooling, the universe enters a matter-dominated phase. The phase plane for a universe model containing such a time interval field is displayed in Figure 2.2. Notice that the inflationary expansion drives Ω closer to unity, rather than away from it as was the case in the more conventional models described in the previous section. The basic argument now is that inflation tends to drive Ω to unity so that, if the inflationary interval lasts long enough, Ω ends up arbitrarily close to unity. At the end of

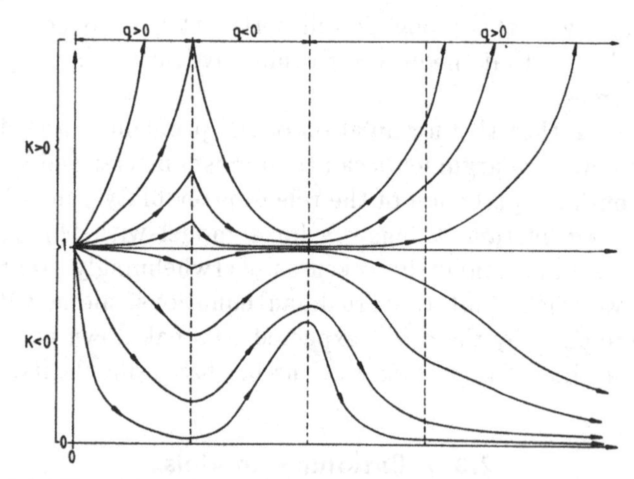

Fig. 2.2 The phase planes for Friedman models with $\gamma < 2/3$ and hence $q < 0$ for some period. From Madsen & Ellis (1988).

inflation Ω drifts away from unity again, but it can remain close to 1 for a very long time. The implication is taken to be that today $\Omega_0 = 1 \pm 10^{-N}$ where $N > 4$ (sometimes $N \gg 4$). The claim, as discussed above, is that if inflation did not take place, considerable 'fine tuning' is required to produce the universe we see today; in particular Ω had to be extremely close to unity at the quantum–classical transition (in order that $|\Omega_0| \leq 1$ today, we require $|\Omega_P - 1| < 10^{-60}$ at the Planck time), which is assumed to be unlikely. Inflation driven by a scalar field or by some effectively equivalent modification of the Einstein field equations removes the need for this fine-tuning at the Planck time. It also does two other important things: it provides a solution to the 'horizon problem', thereby allowing the possibility of physical mechanisms that 'explain' the overall smoothness of the universe on a large scale; and it gives a source for the seed fluctuations to drive gravitational instability and so gives a mechanism to form large-scale structures in the universe (see Chapter 6).

Although this argument is believed in passionately by those cosmologists coming from the particle physics side, it must be treated with considerable care. There are several reasons for caution (and some of the arguments presented confuse various factors at work, and so do not demonstrate what is claimed).

2.3.2 Can inflation produce a low value of Ω?

It was initially claimed that there can be no inflationary models leading to a low value for Ω_0. However, it was then later shown that there are indeed inflationary models that lead to present-day values of Ω_0 other than unity (e.g. Gott 1982; Lucchin & Matarrese 1985; Ellis 1988). Their existence is demonstrated, for a given scalar field potential $V(\phi)$, by considering the (Ω, a) phase planes for inflationary cosmologies, either with a simple fluid representation of inflation (Hubner & Ehlers 1991; Ellis 1991) or using the full scalar field equations (e.g. Belinsky *et al.* 1985; Ellis, Lyth & Mijic 1991). In fact it is easy to construct such models, although they do not fall into the slow-rolling category normally considered in the inflationary literature. Indeed the point of considering them is to consider the full 'phase space' of possibilities. One way to do this is to run the field equations backwards from the chosen (low-Ω) evolution of the scale factor $a(t)$, to find the scalar field potential $V(\phi)$ and evolution $\phi(t)$ that will give the desired result (Ellis & Madsen 1991).

One still has to check that such models can satisfy the various observational constraints on inflationary models, and this typically requires two separate scalar fields (Ellis, Lyth & Mijic 1991) – one field can in principle do the job but requires a rather complex and artificial shape for the potential. This has recently been investigated in depth in a new consideration of this class of models, with proposals of various specific mechanisms that could cause such behaviour. For example, recent 'bubble' inflationary models (e.g. Ratra & Peebles 1995; Bucher, Goldhaber & Turok 1995; Linde & Mezhlumian 1995) envisage the effective potential becoming very steep for large ϕ, or an inflation field non-minimally coupled to gravity. In either case there can be many almost-RW bubbles occurring with various values of the RW parameters, and including any value of Ω_0 between 0 and 1. The virtue of these more recent models is that they can produce models with genuinely negative spatial curvature while at the same time producing a universe which is globally homogeneous. This is difficult in more conventional open universe models where the universe may be expected to be inhomogeneous beyond the curvature scale.

2.3.3 Probability arguments

Given the existence of these models, the argument therefore has again to be phrased in terms of some well-defined invariant measure of the probability of the various universe models. As we discussed above, we do not at present have such a measure, the 'obvious' choice (constant with respect to Ω), giving results that are extremely time-dependent: for example a low-density universe today comes from values of Ω very close to zero at the beginning of inflation but very close to unity at the Planck time-scale. The first problem is that this measure is not invariant as the universe evolves; the least informative measure (2.1) is invariant in this sense, but this result only holds for the simplest matter-dominated cosmologies. Furthermore, there are various different ways one can use a measure once it has been identified. Because the inflationary argument has been given such weight, we shall look at the related issue of measure in in some detail.

Different ways of using probability

There are at least three ways one can use probability arguments in the inflationary universe context, and they produce very different perspectives (Ellis 1991).

The first and most common way is to think of given initial conditions, say $\Omega_{\rm P}$ at $t_{\rm P}$, and the resultant values of Ω at the beginning of inflation ($\Omega_{\rm i}$), at the end of inflation ($\Omega_{\rm f}$), and today (Ω_0). The effect of inflation is considered as follows: if it has not succeeded in producing an $\Omega_{\rm f}$ and Ω_0 very close to 1, this is because inflation did not last long enough. Let it run a little longer, and Ω_0 will be as close to 1 as you like. This can easily be achieved by having a flatter potential or a slightly different coupling constant. Here the perceived fine-tuning of initial conditions is avoided, but it simply replaced by a fine-tuning of the space of functional forms for the inflationary potential. Little seems to have been gained.

The problem with this approach is that effectively it considers probabilities when one varies the physics with fixed initial conditions. We do not regard this as the proper setting for the discussion. Rather than to evaluate the probability of various values of Ω_0 as one varies the physics, one should consider fixed physics and varying conditions at t_i or varying times of observation t_0. In the real universe, at least in the classical epoch we are considering for the purposes of estimating Ω_0 today, one should select which physics is supposed to be in operation and then work out the effect of that chosen set of physical laws. We can, as a separate project, consider the effect of altering the laws of physics; for each chosen set of physical laws, there will be a whole lot of different solutions corresponding to setting different initial conditions for the universe governed by those laws.

Thus the second way is to keep the physics and the initial conditions fixed, and consider the value of Ω_0 as the time of observation t_0 is varied. This is a legitimate and useful exercise; the primary thing one learns here – this can be seen directly from the graphs of Ω vs $a(t)$ – is that when the universe is observed at an arbitrary time, it is very *unlikely* that $\Omega_0 \simeq 1$; rather if $k = -1$ in a particular universe model, the most likely range of Ω to be observed is close to 0, whereas if $\kappa > 0$ all values greater than 1 are approximately equally likely. That is, if we do indeed measure $\Omega_0 \simeq 1$, that is an unstable situation and will not remain true in the future. It can only be a valid prediction for a very restricted time†.

The third context, and the one that is most useful, is to consider given physics and a given time of observation, and to see the implications as we vary initial conditions at either t_P or t_i. The answer depends on the measure of probability chosen for Ω_P or Ω_i.

Different probability measures

In fact the prediction of a critical density today, with or without inflation, has to rely on a probability measure that somehow restricts high density universes at the end of the Planck era (or the start of inflation) to values relatively close to unity – or else it predicts a value greater than unity for Ω_0 today as the most probable value (because arbitrarily high values of Ω are possible at both of those times). The introduction of inflation enables a

† We are assuming here that we do not have the exactly flat case $\kappa \equiv 0$: that will be considered below.

much larger domain of possibilities at that time that is consistent with observations today than if inflation does not occur, but the principle remains the same. It is not obvious what is the 'natural' measure to choose when Planck era effects are taken into account – nor is it clear what it will specify for values of Ω less than the critical value (but greater than zero; we assume $\Omega > 0$).

Quantum theory considerations do suggest one line of thinking†. The argument based on Jaynes' principle we discussed above requires one to accept the Friedman equations as the correct 'laws of physics', based on which we assign the prior measure for Ω. Suppose we abandon these equations and look at the even more poorly specified problem of assigning a measure for the density when all we know is that this density should be positive. We do not even have a dynamical equation for its evolution.

One of the main features of standard Lebesgue measure dx on the real line R is that it is translation invariant: i.e. invariant under the action of the additive group of R on itself. This means that, for example, in wave mechanics the action on wave functions defines a unitary representation of R. When we come to the positive real numbers R_+ the natural group action is the multiplicative action of R_+ on itself. The natural invariant measure here is then not dx but dx/x. Thus, in the wave mechanics of a particle on a positive real line, the action

$$(V(t)\psi)(s) := \psi(st) \tag{2.2}$$

defines a unitary representation of R_+. Of course, the exponential map $\exp : R \to R_+$ associates these two things together, but the difference is still significant. Put another way, when using variables that range from 0 to ∞ one should always consider taking the logarithm of the variable as the basis for a uniform measure. This trick is familiar in general relativity; it is also the route to the well-known Jeffrey's prior in the much wider context of inferential physics (Jeffreys 1939; Jaynes 1969).

If, for example, we apply these comments to Ω at the end of the Planck era, when $a = a_P$ (the Planck scale), then we should adopt the measure $dp = d\Omega/\Omega$. This has the effect of re-scaling the finite range of Ω between 0 and 1, to an infinite range of probability – and making the very low density universes more probable (in this

† We thank C. Isham for these remarks.

natural measure) than those close to unity. This would appear to be a good suggestion for a natural measure – more general than (2.1) because it does not rely on the Friedman equations – but this idea clearly needs much further development.

Alternatively, one can introduce natural measures for the inflationary era and before by (a) examining the Liouville measure for the Hamiltonian system of an FRW model with scalar field (Gibbons, Hawking & Stewart 1987; Hawking & Page 1988), or (b) using the Wheeler–de Witt dynamical metric (Cho & Kantowski 1994; see also Coule 1995). In each case we find first that the resultant probability measure indicates that $\Omega \simeq 1$ is highly preferred, and second that the probability diverges when one integrates over all phase space, so the result is in fact inconclusive. Nevertheless the fact that this measure prefers values near unity 'confirms' the results of our previous simplistic considerations: in the classical case, the fact that Ω is very close indeed to unity at very early times is not the result of fine-tuning, it is dictated by dynamical equations of the system and so is inevitable. These measures are just confirming this result which – as stated in the last section – says that 'fine-tuning' is an illusory problem; when the quantum equations emerge to give a classical limit they should give the same results as the classical equations at that time, which necessarily imply values of Ω very close to 1. This cannot be avoided if we stick to the standard dynamical equations.

The applicability of probability ideas

Suppose we could agree on a solution to these problems and determine a natural measure that predicts a value of Ω at the Planck time, leading to prediction of a range of values for Ω_0 today based on that measure. Then it is still not clear that its use makes sense in the cosmological context, no matter how strongly physicists feel about it. The essential point here is that major stumbling block in all of cosmology: the uniqueness of the universe. This uniqueness leads one away from the usual 'frequentist' notions of probability and towards the Bayesian interpretation, which we discussed briefly above and in which the probabilities of unique events can be handled without conceptual problems (e.g. Garrett & Coles 1993). In practice, however, the need to incorporate all prior information and posterior selection effects in the Bayesian analysis poses formidable technical problems. We are still a long way away from being able to make meaningful statements about which kind of universe is more probable than another.

We could of course try to claim that quantum cosmology provides a unique specification – if we could overcome all the technical and conceptual problems of that subject, and then obtain a unique answer for the state of the universe predicted at the end of the Planck era. However, disregarding the technical problems, there are various competing methods for specifying how to determine the initial state of the universe from quantum cosmology considerations – so the required uniqueness remains elusive.

2.3.4 The modification of Ω by inflation

Finally on this matter, we reiterate that there are classes of universes (power-law inflation, the coasting universe families) that are essentially inflationary but do *not* drive Ω to unity as standard inflation does. Indeed in the coasting solutions – which are consistent with some superstring-motivated cosmologies as well as some string-dominated models – Ω stays constant during the coasting period. This is because they effectively have a value of $\gamma = 2/3$. Although marginal in terms of inflation, this will still solve the horizon and monopole problems (Linde 1990; Ellis, Lyth & Mijic 1991; Coles & Lucchin 1995). There is no fundamental reason why the early universe should not have evolved in this way rather than the more commonly supposed 'slow-rolling' standard inflation. Thus the conclusion that inflation necessarily drives Ω towards 1 from its initial value is true for some but not all inflationary models.

Some would argue that these marginal models were less likely than the standard ones, but this merely takes us back to another variation of the probability problem and is not a serious objection given all that has come before!

2.3.5 The occurrence of inflation

We realise that these issues are controversial. Thus while we believe they have merit, we will not hang our case on them. In the end the main point as regards inflation is simple: despite some dogmatic statements that have been made from time to time, *we do not in fact have any proof that inflation took place in the early universe, nor do we even have any well-defined candidate for the 'inflaton' scalar field.* Thus, even if all the previous comments are discounted, the argument from inflation does not by itself lead one to conclude without doubt that $\Omega_0 = 1$ in the real universe – for it may not have happened. The real situation is the other way round: the critical astrophysicist is most likely to be persuaded that inflation did indeed take place if we do succeed in showing from observations of the universe that $\Omega_0 = 1$ to high accuracy. But this too is not a watertight argument, for $\Omega_0 = 1$ could have been produced without inflation as well, and indeed is favoured by the least-informative measure we have presented above. Nevertheless we recognise its merit. For present purposes the point is that whatever attitude one may take regarding the probability of inflation driving Ω to 1, inflation may not have taken place; thus the argument from inflation cannot be taken as conclusively showing $\Omega_0 = 1$ today.

2.3.6 The exactly critical case

In all the above and in what follows, statements such as that in the last paragraph ('$\Omega_0 = 1$ today') do *not* mean $\Omega_0 = 1$ *exactly*, but rather to very high approximation: say $\Omega_0 = 1 \pm 10^{-4}$ or better. The reason for this in the context of inflation is that inflation itself produce fluctuations in the matter of that order, so Ω_0 can never be defined with more precision than this. It has never been part of the inflationary story that inflation implies $\Omega_0 = 1$ exactly. Indeed if this were so, we would have the critical (Einstein–de Sitter, $\kappa = 0$) case; then at *all* times in the history of the universe $\Omega = 1$ exactly, and inflation makes no difference whatever to the value of Ω, which always remains at that critical value. Thus this model is unstable in the same sense as the Einstein static universe is unstable: initial data would have to be set infinitely accurately to achieve this universe; quantum fluctuations would prevent this accuracy. Physically, it cannot be achieved. In terms of mathematical modeling, when one takes into account the lumpy nature of the real universe (see Chapter 8 below), it hardly makes sense to suggest the best-fit background universe model could have *exactly* the critical value, which requires *infinite* precision in terms of the value of Ω_0: we are then claiming that the best-fit model cannot have a non-zero number in the 157th, 1089th, one millionth, or any other decimal place in the quantity $\Omega - 1$. We only make these comments because misunderstandings about it have not yet been dispelled (e.g. Binney 1993).

2.4 The cosmological constant

The standard argument from inflation basically leads one to infer that the spatial sections should be flat. As is obvious from equation (1.15), this does not require that $\Omega_0 = 1$ unless we simultaneously require $\Lambda = 0$. The cosmological constant term is not obligatory in inflation, but neither is it excluded by the theory. As recently discussed by Efstathiou (1995), there are two basic arguments (independent of those for and against inflation) that might lead one to include such a term: the possible discrepancy between the ages of the oldest stars in globular clusters and the age of the universe inferred from Friedman models with $\Lambda = 0$ (which we discuss in Chapter 3), and observations of large-scale structure which exclude flat universe models based on cold relic particles. The nature of cold dark matter is discussed below, in §2.5, and the galaxy clustering results are discussed in Chapter 6.

2.4.1 Flatness without a critical density

From equation (1.15) we can infer that the value required for Λ in order to guarantee flat spatial sections is

$$\Lambda = 3H_0^2(1 - \Omega_0). \tag{2.3}$$

It is often convenient to express Λ in terms of the critical density by introducing

$$\Omega_\Lambda \equiv \frac{\Lambda}{3H_0^2}, \tag{2.4}$$

a notation which reinforces the fact that the cosmological constant does correspond to a vacuum energy density. As we shall discuss in Chapter 3, models described by (2.3) are in principle distinguishable from $\Omega = 1$ models by the difference in their expansion rate encoded in the deceleration parameter q_0 and their effects on gravitational lensing. Notice also that the relative contributions of Λ and Ω to the dynamics evolve with time: the effect of Λ is negligible at early times, but dominates at late times.

2.4.2 Fine-tuning of Λ

Most of the problems we raised in connection with inflationary models with $\Lambda = 0$ are also pertinent to this model. Indeed, in some respects, one has the 'coincidence' argument twice over. For a start, one needs to have a vacuum density of the right order of magnitude and then, after that, one needs to accept that we live at the epoch when the cosmological constant term is just beginning to dominate over the effect of matter. The second of these problems is similar to those we have discussed above. The problem with the first of these considerations is as follows.

A cosmological Λ term behaves like a fluid with equation of state of the form

$$\rho_\Lambda = -\frac{p_\Lambda}{c^2} \equiv \frac{\Lambda c^2}{8\pi G}, \tag{2.5}$$

so that from the Friedman equation we get

$$\frac{\kappa}{a_0^2} = \frac{H_0^2}{c^2}(\Omega_0 + \Omega_\Lambda - 1). \tag{2.6}$$

A very crude observational limit on Ω_Λ can be imposed by simply requiring that it does not exceed critical density by, say, a factor two; stronger limits from gravitational lensing frequencies are discussed in Chapter 3. The result is

$$|\rho_\Lambda| < 2\rho_{\rm cr} \simeq 4 \times 10^{-29}\ {\rm g\ cm^{-3}} \simeq 10^{-46}\frac{m_{\rm n}^4}{(\hbar/c)^3} \simeq 10^{-48}\ {\rm GeV}^4, \tag{2.7}$$

which has been referred to a typical particle physics energy scale for reference ($m_{\rm n}$ is the mass of a nucleon; we have used 'natural' units in which $\hbar = c = 1$). This relation corresponds to

$$|\Lambda| < 10^{-55}\ {\rm cm^{-2}}. \tag{2.8}$$

From Λ one can also construct a quantity which has the dimensions of a mass:

$$m_\Lambda = \left[|\rho_\Lambda|\left(\frac{\hbar}{c}\right)^3\right]^{1/4} = \left(\frac{\hbar^3}{8\pi G c}|\Lambda|\right)^{1/4} < 10^{-32}\ {\rm eV}. \tag{2.9}$$

One can illustrate the severity of even this crude constraint by comparing m_Λ with the upper limit on the mass of the photon: according to recent estimates this is $m_\gamma < 3\times 10^{-27}$ eV. The problem of the cosmological constant lies in the fact that the quantities $|\Lambda|$, $|\rho_\Lambda|$ and $|m_\Lambda|$ are so extremely and, apparently, 'unnaturally' small.

A more modern interpretation of Λ is that ρ_Λ and p_Λ represent the density and pressure of the *vacuum*, which is understood to be analogous to the ground state of a quantum system (the equation of state $p_v = -\rho_v c^2$ comes from the Lorentz-invariance of the energy-momentum tensor of the vacuum). In modern theories of elementary particles with spontaneous symmetry breaking it turns out that

$$\rho_v \simeq V(\Phi, T). \tag{2.10}$$

Modern gauge theories predict that

$$\rho_v \simeq \frac{m^4}{(\hbar/c)^3} + \text{ constant}, \tag{2.11}$$

where m is the energy at which the symmetry-breaking transition occurs (10^{15} GeV for GUT transitions, 10^2 GeV for the electroweak transition, 10^{-1} GeV for the quark–hadron transition and (perhaps) 10^3 GeV for a supersymmetric transition). In the symmetry-breaking phase one has a decrease of ρ_v of order

$$\Delta\rho_v \simeq \frac{m^4}{(\hbar/c)^3}, \tag{2.12}$$

corresponding to 10^{60} GeV^4 for the GUT, 10^{12} GeV^4 for supersymmetry, 10^8 GeV^4 for the electroweak transition, and 10^{-4} GeV^4 for QCD.

In the light of these comments the cosmological constant problem can be posed in a clearer form:

$$\begin{aligned} \rho_v(t_P) &= \rho_v(t_0) + \sum_i \Delta\rho_v(m_i) \simeq 10^{-48}\text{GeV}^4 + 10^{60}\text{ GeV}^4 \\ &= \sum_i \Delta\rho_v(m_i)(1 + 10^{-108}), \end{aligned} \tag{2.13}$$

where $\rho_v(t_P)$ and $\rho_v(t_0)$ are the vacuum density at the Planck and present times respectively and m_i represents the energies of the various phase transitions which occur between t_P and t_0. This equation can be phrased in two ways: $\rho_v(t_P)$ must differ from

$\sum_i \Delta\rho_v(m_i)$ over the successive phase transitions by only one part in 10^{108} or the sum $\sum_i \Delta\rho_v(m_i)$ must, in some way, arrange itself so as to satisfy (2.13). Either way, there is definitely a problem of extreme 'fine-tuning' in terms of $\rho_v(t_P)$ or $\sum_i \Delta\rho_v(m_i)$.

At the moment, there exist only a few theoretical models which even attempt to resolve the problem of the cosmological constant (e.g. Coleman 1988). Indeed, many cosmologists regard this problem as the most serious one in all cosmology (e.g. Weinberg 1989). This is strictly connected with the theory of particle physics and, in some way, to quantum gravity. Inflation, in particular, does not solve this problem.

A possible way to avoid this problem has been discussed by Efstathiou (1995). In an inflationary scenario in which one envisages an ensemble of universes each with different values of Ω and Λ, one can make plausible arguments as to why those 'bubbles' with $\Omega_0 \simeq \Omega_\Lambda$ might be more efficient at producing galaxies and might, therefore, be favoured on anthropic grounds. This explanation is, of course, highly speculative but it does illustrate the importance of taking anthropic selection effects into account.

2.5 Particle physics

We now turn to arguments from particle physics theory about the nature and quantity of relic particles that may have been produced in the big bang. This class of argument differs from the inflationary argument in that inflation does not really predict what form the matter has to be, even if the particular model in question does predict that $\Omega_0 = 1$. Regardless of inflation, and if the universe is indeed dominated by non-baryonic matter, it is also obviously important to figure out the present density of various types of candidate particle expected to be produced in the early stages of the big bang. For a list of several hypothetical such relics, see for example Trimble (1987).

In general, we shall use the suffix X to denote some generic particle species produced in the early universe; we call such particles *cosmic relics.* We distinguish at the outset between two types of cosmic relics: *thermal* and *non-thermal.* Thermal relics are held in thermal equilibrium with the other components of the universe until they decouple; a good example of this type of relic is the massless neutrino, although this is of course not a candidate for the gravitating dark matter. One can subdivide this class into *hot* and *cold* relics. These are usually labelled as hot dark matter (HDM) and cold dark matter (CDM) respectively. The former are relativistic when they decouple, and the latter are non-relativistic. Non-thermal relics are not produced in thermal equilibrium with the rest of the universe. Examples of this type would be monopoles, axions and cosmic strings. The case of non-thermal relics is much more complicated than the thermal case, and no general prescription exists for calculating their present abundance. We shall therefore concentrate here on thermal relics, which seem to be based on better established physics (Cowsik & McClelland 1972; Lee & Weinberg 1977), and for which a general treatment is possible. In practice, it turns out in fact that this approach is also quite accurate for particles like the axion anyway (e.g. Coles & Lucchin 1995).

The time evolution of the number density n_X of some type of particle species X is generally described by the Boltzmann equation:

$$\frac{\mathrm{d}n_X}{\mathrm{d}t} + 3\frac{\dot{a}}{a}n_X + \langle\sigma_A v\rangle n_X^2 - \psi = 0, \tag{2.14}$$

where the term in $\dot{a}/a$ takes account of the diluting effect of the expansion of the universe, $\langle\sigma_A v\rangle n_X^2$ is the rate of collisional annihilation (σ_A is the cross-section for annihilation reactions, and v is the mean particle velocity); ψ denotes the rate of creation of particle pairs. If the creation and annihilation processes are negligible, one has the expected solution: $n_{X eq} \propto a^{-3}$. This solution also holds if the creation and annihilation terms are non-zero, but equal to each other, i.e. if the system is in equilibrium: $\psi = n_{X eq}^2\langle\sigma_A v\rangle$. Thus, equation (2.14) can be written in the form

$$\frac{\mathrm{d}n_X}{\mathrm{d}t} + 3\frac{\dot{a}}{a}n_X + \langle\sigma_A v\rangle(n_X^2 - n_{X eq}^2) = 0 \tag{2.15}$$

or, if we introduce the comoving density

$$n_c = n\left(\frac{a}{a_0}\right)^3, \tag{2.16}$$

in the form

$$\frac{a}{n_{c,eq}}\frac{\mathrm{d}n_c}{\mathrm{d}a} = -\frac{\tau_H}{\tau_{\mathrm{coll}}}\left[\left(\frac{n_c}{n_{c,eq}}\right)^2 - 1\right], \tag{2.17}$$

where $\tau_{\rm coll} = 1/\langle\sigma_A v\rangle n_{eq}$ is the mean time between collisions and $\tau_H = a/\dot{a}$ is the characteristic time for the expansion of the universe; we have dropped the subscript X for clarity. Equation (2.17) has the approximate solution

$$\begin{aligned} n_c &\simeq n_{c,eq}, \\ n_c &\simeq \text{constant} \simeq n_c(t_d) \end{aligned} \tag{2.18}$$

in cases where $\tau_{\rm coll} \ll \tau_H$ and vice versa respectively; t_d is the moment of 'freezing out' of the creation and annihilation reactions, defined by

$$\tau_{\rm coll}(t_d) \simeq \tau_H(t_d). \tag{2.19}$$

2.5.1 Hot thermal relics (HDM)

As we have explained, hot thermal relics are those that decouple while they are still relativistic. We basically have light neutrinos in mind here. Let us assume that the particle species X becomes non-relativistic at some time t_{nX}, such that $Ak_BT(t_{nX}) \simeq m_X c^2$ ($A \simeq 3.1$ or 2.7 is a statistical-mechanical factor which takes these two values according to whether X is a fermion or a boson). For simplicity we take $A = 3$ to get rough estimates. Hot relics are thus those for which $t_{nX} > t_{dX}$, where t_{dX} is defined above.

Let us denote by g_X the statistical weight of the particle X and by g_X^* the effective number of degrees of freedom of the universe at t_{dX}. Applying the rule of conservation of entropy per unit comoving volume, we have

$$g_X^* T_{0X}^3 = 2T_{0r}^3 + \frac{7}{8} \times 2 \times N_\nu\, T_{0\nu}^3 = g_0^* T_{0r}^3, \tag{2.20}$$

where T_{0X} is the present value of the effective temperature defined by the mean particle momentum via $\overline{p}_X \simeq 3k_BT_X/c$ and T_{0r} is the present temperature of the photon background and $T_{0\nu} = (4/11)^{1/3}T_{0r}$ takes account of the N_ν neutrino families; $g_0^* \simeq 3.9$ for $N_\nu = 3$. We thus obtain

$$T_{0X} = \left(\frac{g_0^*}{g_X^*}\right)^{1/3} T_{0r}. \tag{2.21}$$

This equation also applies to neutrinos if one puts

$$g_\nu^* = 2 + \frac{7}{8} \times 2 \times N_\nu + \frac{7}{8} \times 2 \times 2 \tag{2.22}$$

(photons, neutrinos and electrons all contribute to g_ν^*). In this case we obtain the well-known relation

$$T_{0\nu} = \left(\frac{4}{11}\right)^{1/3} T_{0r} = 0.7\, T_{0r}. \qquad (2.23)$$

The present number-density of X particles is then

$$n_{0X} \simeq 0.5 B g_X \left(\frac{T_{0X}}{T_{0r}}\right)^3 n_{0r} \simeq 0.5 B g_X \frac{g_0^*}{g_X^*} n_{0r}, \qquad (2.24)$$

where $B = 3/4$ or 1 according to whether the particle X is a fermion or a boson. The density parameter corresponding to these particles is then just

$$\Omega_X = \frac{m_X n_{0X}}{\rho_{0c}} \simeq 2 B g_X \frac{g_0^*}{g_X^*} \frac{m_X}{10^2\ \mathrm{eV} h^2}\ . \qquad (2.25)$$

To give an illustrative example, consider hypothetical particles with mass $m_X \simeq 1$ keV, which decouple at $T \simeq 10^2$ to 10^3 MeV when $g_X^* \simeq 10^2$; these have $\Omega_X \simeq 1$.

Let us now apply equation (2.25) to the case of a single massive neutrino species with $m_\nu \simeq 1$ MeV, which decouples at a temperature of a few MeV when $g_X^* = 10.75$ (taking account of photons, electrons and three types of massless neutrinos). The condition that the cosmic density of such relics should not be much greater than the critical density requires that $m_\nu <$ 90 eV: this bound was obtained by Cowsik & McClelland (1972). If, instead, all the neutrino types have masses around 10 eV, then their contribution to the critical density will be given by

$$\Omega_\nu h^2 \simeq 0.1 N_\nu \frac{\langle m_\nu \rangle}{10\ \mathrm{eV}}\ , \qquad (2.26)$$

where $\langle m_\nu \rangle$ is the average neutrino mass if we allow more than one species to have non-zero rest mass. As we do not know this mass, we do not obtain any reliable estimate of the contribution to Ω_0 from such particles. However, the physics governing the production of neutrinos in the early universe is sufficiently well-understood that a laboratory measurement of a non-zero neutrino mass would be accepted by most astrophysicists as necessarily implying the corresponding contribution to Ω_0.

2.5.2 Cold thermal relics (CDM)

Calculating the density of cold thermal relics is much more complicated than for hot relics. At the moment of their decoupling the number density of particles in this case is given by a Boltzmann distribution:

$$n(t_{dX}) = g_X \frac{1}{\hbar^3}\left(\frac{m_X k_B T_{dX}}{2\pi}\right)^{3/2} \exp\left(-\frac{m_X c^2}{k_B T_{dX}}\right). \qquad (2.27)$$

The present density of cold relics is therefore

$$n_{0X} = n(t_{dX})\left[\frac{a(t_{dX})}{a_0}\right]^3 = n(t_{dX})\frac{g_0^*}{g_X^*}\left(\frac{T_{0r}}{T_{dX}}\right)^3. \qquad (2.28)$$

The problem in this case is to find T_{dX}, that is to say the temperature at which the particles decouple. The characteristic time for collisional annihilations is given by

$$\tau_{\rm coll}(t_{dX}) = \left[n(t_{dX})\sigma_0\left(\frac{k_B T_{dX}}{m_X c^2}\right)^q\right]^{-1}, \qquad (2.29)$$

where we have made the assumption that

$$\langle \sigma_A v \rangle = \sigma_0 \left(\frac{k_B T}{m_X c^2}\right)^q : \qquad (2.30)$$

$q = 0$ or 1 for most kinds of reaction. Introducing the variable $x = m_X c^2/k_B T$, the condition $\tau_{\rm coll}(x) = \tau_H(x)$ is true when $x = x_{dX} = m_X c^2/k_B T_{dX} \gg 1$. An approximate solution of the condition for decoupling (2.19) then yields the present density of relic particles:

$$\rho_{0X} \simeq 10 g_X^{*\,-1/2} \frac{(k_B T_{0r})^3}{\hbar c^4 \sigma_0 m_P} x_{dX}^{n+1}; \qquad (2.31)$$

where we have referred the mass scale to m_P, the Planck mass.

As an application of this equation, one can consider the case of a heavy neutrino of mass $m_\nu \gg 1$ MeV. If the neutrino is a Dirac particle (i.e. if the particle and its antiparticle are not equivalent) then the cross-section in the non-relativistic limit varies as v^{-1}, corresponding to $q = 0$, for which $\sigma_0 = \text{constant} \simeq 0.8 g_{\rm wk}^2 m_\nu^2 c/\hbar^4$ ($g_{\rm wk}$ is the weak interaction coupling constant). Putting $g_\nu = 2$ and $g_\nu^* \simeq 60$ one finds that $x_{d\nu} \simeq 15$, corresponding to a temperature $T_{d\nu} \simeq 70(m_\nu/\text{GeV})$ MeV. If we place this value of $x_{d\nu}$ in equation (13.4.7), the condition that $\Omega_\nu h^2 < 1$ implies that $m_\nu > 1$ GeV: this limit was found by Lee & Weinberg (1977). If, on the other hand, the neutrino is a Majorana particle (i.e. if the particle and its antiparticle are equivalent), the annihilation rate $\langle\sigma_A v\rangle$ has terms in x^{-q} with $q = 0$ and 1, thus complicating matters considerably. Nevertheless, the limit on m_ν we found above does not change. In fact we find $m_\nu > 5$ GeV. If the neutrino has mass $m_\nu \simeq 100$ GeV, the energy scale of the electroweak phase transition, the cross-section is of the form $\sigma_A \propto T^{-2}$ and all the previous calculations must be modified.

As in the previous case, we do not know the relevant masses, and so do not obtain any reliable estimate of the contribution to Ω_0 from such particles.

2.5.3 Decaying dark matter

There is a further line of theoretical argument for a high value of Ω_0 which, in contrast to the others presented in this chapter, is essentially astrophysically based. The decaying neutrino theory of Sciama (1990, 1993) leads to rather definite values of both Ω_0 and H_0. In this theory, massive dark matter neutrinos decay into photons and less massive neutrinos. The photons are assumed to be the main ionising agent for hydrogen in galaxies and the intergalactic medium. Various astronomical arguments then lead to a well-defined value for the decay photon energy of 14.6 ± 0.1 eV, and a decay lifetime of $1\text{–}3 \times 10^{23}$ s. This theory makes contact with a number of different astronomical and cosmological phenomena, and agrees well with a variety of observations; see Sciama (1993) for a review. The theory leads to well-defined values of Ω_0 and H_0 in the following manner. If the residual neutrino in the decay has much less mass than the original neutrino (as is likely on a number of grounds), the decaying neutrino has a mass which is twice the energy of the decay photon, that is, 29.2 ± 0.2 eV. Since the cosmological number density of neutrinos of each type is fixed at $3/11$ times the photon density of the cosmic microwave background, one knows $\Omega_\nu h^2$. Adding in the (small) baryonic contribution, from the considerations of primordial nucleosynthesis we discuss in Chapter 4, one derives $\Omega_0 h^2 \simeq 0.31 \pm 0.06$. Assuming that $\Lambda = 0$ and that the age of the universe exceeds 12 billion years, one finds that $\Omega_0 \geq 1$. If Ω_0 is exactly unity, one deduces that $H_0 = 56 \pm 0.5$ km s^{-1} Mpc^{-1}, and an age of 12 billion years; see Chapter 3. This is indeed an intriguing line of argument, relating a whole series of astronomical observations to plausible particle physics processes, and in line with the suggestion of Turner, Steigman & Krauss (1984) as to how to reconcile the theoretical prejudice for $\Omega_0 = 1$ with the astronomical data.

However, so far the required massive neutrino has not been detected. Furthermore the value required for h is not in line with the observational trends discussed in the next chapter. Thus we note this as an interesting proposal to keep in mind, which may yet receive experimental confirmation through detection of the required massive neutrinos, but is not binding at present.

2.5.4 Possible laboratory tests?

Aside from their possible cosmological consequences, which we shall discuss in subsequent chapters but which are in any case indirect, the only ways to constrain the form of particle dark matter is through local astrophysical measurements or laboratory tests.

Weakly interacting particles with masses greater than a few GeV would have sufficiently large cross-sections for interactions with nucleons that they could become trapped in the Sun (Spergel & Press 1985). These would then interfere with energy transport in the solar core and a constraint on the allowed particle masses thus emerges which excludes, for example, a significant abundance of Dirac neutrinos with mass exceeding 6 GeV. If our halo were comprised of weakly interacting massive particles (WIMPs) then there is a possibility that they could be detected in laboratory experiments. When a WIMP collides with an atomic nucleus, there is expected to be a recoil. If the nucleus were embedded in a very cold solid then the rise in temperature of the solid might be just detectable, either using superconducting grains held near their critical temperature or by attempting to detect the 'phonons' produced by individual recoil events (Smith 1986). If the dark matter consisted of axions, then these could in principle be detected by conversion into photons in the presence of strong magnetic fields. The axion mass is poorly known, however, and the energy range of the photons produced is consequently highly uncertain, so this appears a less promising possibility (Sikivie 1985).

So far, many brave attempts have been made to detect these hypothetical particles but they have all proved fruitless; for a review of recent developments in this field, see Spergel (1996). If they succeed, they will be fundamentally important. Until then, they are merely a promising line of attack that has yet to yield results. We shall not, therefore, discuss the possibilities of direct detection among the other observational evidence we consider in later chapters.

2.6 Summary

The various theoretical arguments are suggestive but not conclusive. They link the cosmological issue to the rest of physics in a deep manner, however at the same time extending standard physics way beyond the area where it has been tested.

Furthermore the issue of initial conditions enters in an essential way, and is not covered by any normal physical theory. We can speculate as to what might have happened and what probability principles may have been in operation, but it is essential to realise that this is the precise situation – these are and will remain speculations. Hence we end this chapter as we began it, by pointing out that any of these theoretical arguments will be validated if and only if observations support their conclusions.

3
Cosmological observations

In this chapter we shall explore the evidence emerging from bulk properties of the universe. This in effect boils down to two kinds of arguments, those based around the age of the universe and those pertaining directly to its spatial geometry. A general review of the first kind of argument is given by Hendry & Tayler (1996) while the second kind is discussed in more detail by Sandage (1988, 1995).

3.1 The age of the universe

The idea behind the use of ages to test cosmological models is very simple: there should be nothing in the universe that is older than the universe. The oldest observed astronomical object therefore gives a lower limit on the age of the universe. Given present-day values of Ω_0, H_0 and Λ one can solve the Friedman equation to find t_0, the time elapsed since $t = 0$ when $a(t) = 0$.

3.1.1 Model ages

The form of the Hubble expansion law, equation (1.7), dictates that the characteristic time for the expansion is defined by the Hubble time, $\tau = 1/H_0$. Notice, however, that in models dominated by 'normal' matter the expansion is decelerating because of the gravitational attraction of the mass contained within the universe. The actual age of the universe is therefore expected to be less than the Hubble time. In fact, it is straightforward to show that, for matter-dominated universes,

$$t_0 = f(\Omega_0)\tau, \tag{3.1}$$

where the function f is relatively slowly varying and, for the case $0 < \Omega_0 < 1$, which is most relevant here, we get

$$f(\Omega_0) = (1 - \Omega_0)^{-1} - \frac{\Omega_0}{2}(1 - \Omega_0)^{-3/2} \cosh^{-1}\left(\frac{2}{\Omega_0} - 1\right), \quad (3.2)$$

which is less than 1. For example, $f = 2/3$ for $\Omega_0 = 1$ (the Einstein–de Sitter universe) and $f = 1$ for $\Omega_0 = 0$; for $\Omega_0 = 0.1$, $t_0 \simeq 0.9\tau$. An old universe therefore favours a low Ω_0 (for a fixed value of H_0).

In the presence of a cosmological constant, the relation (3.2) becomes more complicated and, in general, there is no simple equation relating Ω_0, Ω_Λ and t_0. A closed-form expression is, however, available for the 'inflationary' $k = 0$ models† containing a cosmological constant:

$$t_0 = \frac{2}{3H_0}\left[\frac{1}{2\sqrt{1-\Omega_0}} \log \frac{1+\sqrt{1-\Omega_0}}{1-\sqrt{1-\Omega_0}}\right]. \quad (3.3)$$

Generally speaking, however, one can see that a positive cosmological constant term tends to act in the direction of accelerating the universe and therefore tends to increase the age relative to decelerated models; indeed $f > 1$ for sufficiently large Λ.

To summarise, then, we find that matter-dominated models with $\Lambda = 0$ have

$$2/3 < H_0 t_0 < 1, \quad (3.4)$$

while $H_0 t_0 > 1$ can only be achieved in the presence of a positive cosmological constant.

3.1.2 The Hubble constant

There is a fundamental dependence on H_0 in all the big-bang models, whether with or without a cosmological constant, so it is important to discuss this parameter before going any further. As we mentioned in Chapter 1, H_0 is not known to any great accuracy and the present state of ignorance is usually represented in the literature by the constant h, where $H_0 = 100\,h\,\mathrm{km\,s^{-1}\,Mpc^{-1}}$.

† The age of the universe is essentially independent of whether there was inflation or not. We are concerned with the time elapsed since decoupling of matter and radiation (which comes after inflation), for this is the time in which macroscopic objects could form.

There has been a broad consensus for many years that h lies in the range $0.4 < h < 0.9$. With this definition of h, the Hubble time is roughly $10h^{-1}$ Gyr so that an Einstein–de Sitter universe has an age of around 13 Gyr if $h \simeq 0.5$ and only 6.7 Gyr if $h \simeq 1$. This result is the reason why most theorists prefer a low value of h and, until recently, there has been enough uncertainty in the observational evidence to accommodate this value without undue difficulty.

There is no space in this book to go into the details of the observational procedures involved in estimating H_0; all the most important techniques are discussed at length by Rowan-Robinson in his book *The Cosmological Distance Ladder* (1985). The main cause of the uncertainty in h has for some time been the inability to calibrate relative distance estimates for distant galaxies against direct distance estimates to nearby galaxies. The way one measures distances to nearby galaxies is by attempting to resolve individual stars, particular Cepheids whose period–luminosity relationship can be used to infer distance from observations of apparent brightness and time-scale of variability. Relative distance indicators of galaxies are generally calibrated to the Virgo cluster of galaxies – the nearest large concentration of galaxies to the Local Group, while Cepheid distances are generally calibrated on the Large Magellanic Cloud. What has been needed, therefore, is a direct measurement of Cepheid distance to Virgo galaxies to extend the distance ladder out to where relative distances between galaxies are reliable. Measuring light curves of individual stars in Virgo galaxies is a very difficult task using ground-based telescopes, but is possible using the Hubble Space Telescope (HST) and was indeed one of the major motivations for the construction of this instrument.

An emerging consensus in favour of $h \simeq 0.8$ (Fukugita, Hogan & Peebles 1993) has to a large extent been confirmed by HST observations of Cepheid distances to Virgo galaxies (Freedman *et al.* 1994) giving $h = 0.8 \pm 0.17$. Ground-based estimates also seem to be consistent with this: Pierce *et al.* (1994) quote $h = 0.87 \pm 0.07$, though most workers agree that the latter authors have substantially underestimated their errors. Subsequent estimates of distance to galaxies in the Leo I cluster (Tanvir *et al.* 1995) give $h = 0.69 \pm 0.08$. It is important to stress that these results are really only preliminary indications based on individual galaxies, and some doubts still remain that Cepheids in the relevant galaxies may differ systematically from those in the LMC to which they are normalised, but it is fair to say that the balance of the evidence is now creeping up towards at least $h \simeq 0.7$ rather than $h \simeq 0.5$. We must wait for the various HST programmes to run their course before making more definitive statements. History also has a lesson to teach us here: it is not so long ago that a value of $h \simeq 2.5$ was the best observational bet (Baade 1956).

The determinations we have mentioned here all rely on the classical distance ladder of interleaving relative distance measurements, founded upon a bottom rung of local geometric measurements (Rowan-Robinson 1985). It is worth mentioning, however, that there is a possibility that this somewhat unwieldy apparatus may become redundant in the relatively near future. Recent progress on the use of Type Ia supernovae as standard candles will bypass much of the distance ladder (see Section 3.2.1 below). Other methods, such as the Sunyaev–Zel'dovich effect in clusters (Jones *et al.* 1993) and time-delay effects in gravitational lens systems (Blandford & Kochanek 1987), may circumvent this approach altogether and give direct geometric estimates of extragalactic distances.

3.1.3 Ages of objects

Given a model, and an observationally inferred value of h, the way to test cosmological models is to compare the predicted age of the universe with the inferred ages of astronomical objects (e.g. Sandage 1982). There are many ways one can do this, at least in principle. One can date lunar and meteoritic rock using long-lived radioactive isotopes such as uranium 235. Since these heavy isotopes were presumably formed in supernova explosions, a little bit of modelling of the rate of such explosions allows one to estimate the age of the Galaxy. Typical numbers quoted for this type of analysis are $t \simeq (9 - 16)$ Gyr, the range in estimates here mainly represents uncertainty in the modelling (Symbalisky & Schramm 1981).

Stellar evolution also provides stringent limits on t_0. The very oldest stars are observed in globular clusters and these objects can be dated by fitting the calculated locus of points traced by stars of differing mass but constant age (isochrones) to the Hertzsprung–Russell diagrams of the stars they contain. The theoretical side of this procedure relies on detailed knowledge of stellar structure (including such troublesome phenomena as convection and chemical mixing) and of physical parameters (such as nuclear reaction rates and opacities). There is, therefore, some uncertainty in the theoretical predictions which is hard to quantify. Ages have been quoted up to 20 Gyr for some of the oldest globular clusters, but many of these claims are still controversial. A recent analysis of the ages of older globular clusters by Bolte & Hogan (1995) concludes that globular clusters in our Galaxy have ages in the range $t_G = 15.8 \pm 2.1$ Gyr, although Chaboyer *et al.* (1996a,b) give arguments that suggest $t_G \simeq 12.8$ may be a more reliable figure for M68, which was previously thought to be one of the oldest such systems†. Using a different method, not based on conventional isochrone fitting, Jimenez *et al.* (1996) derive still lower ages for the oldest systems, around $t_G = 13.5 \pm 2$ Gyr.

Not being experts on stellar evolution, we think a reasonable indication of the possible tolerance in the estimated ages of the oldest globular clusters is given by $12 < t_G < 18$ Gyr. For globular clusters to be significantly younger than this there would have to be a big error in the understanding of stellar evolution which is, of course, not impossible.

Another constraint is supplied by observations of galaxies at high redshift. If an object is detected at a cosmologically significant distance then the look-back time involved may be comparable to the age of the universe. One can attempt to fit the optical emission spectrum of such an object with a synthetic spectrum obtained using a stellar population of some assumed age. A constraint on allowed cosmological models is then imposed by requiring that the age of the universe t_z corresponding to the redshift of the object z should exceed the object's inferred age. The dating of a galaxy in this way seems to be subject to a greater uncertainty than the dating of globular clusters of stars, but recent estimates indicate that it is difficult to accommodate the existence of appar-

† That globular cluster stars may be older than the number suggested by radioactive cosmochronology is not surprising if globular clusters formed a long time before the bulk of the stars in the galactic disk.

ently old galaxies at $z > 1$ in the Einstein–de Sitter model (e.g. Dunlop *et al.* 1996).

3.1.4 Discussion

Even allowing no significant time-lag between the big bang and the onset of globular cluster formation, our interpretation of the globular age estimates would lead one to the conclusion that $t_0 >$ 12 Gyr. This is already in conflict with the Einstein–de Sitter universe with $\Lambda = 0$ unless $h < 0.54$. If it turns out that $h \simeq 0.7$, then we will simply have no alternative but to abandon the idea that $\Omega_0 = 1$. On the other hand, moving to a low value of Ω_0 does not buy very much breathing space: a universe with $\Omega_0 = 0.1$ and $h = 0.5$ has an age of 18 Gyr, comfortably in accord with age estimates; if $h = 0.7$ this reduces to 13 Gyr, so that $h > 0.7$ is also a problem unless Ω_0 is very small indeed. Even an empty universe would be only 16 Gyr old if $h = 0.8$.

Since, as we shall argue, it seems a fairly safe bet that Ω_0 is at least 0.1, a high value of $h > 0.7$ would leave us with the alternatives† of either adding a cosmological constant or abandoning the big-bang model entirely (e.g. Arp *et al.* 1990).

3.1.5 The cosmological constant

The looming crisis apparent in the discrepancy between age estimates and the inverse of the Hubble parameter suggests the possibility that $f > 1$, which, in turn, requires the existence of a cosmological Λ-term. A cosmological constant also allows what could be considered an advantage: consistency of a low-matter-density universe with flat spatial sections (Chapter 2; see also Ostriker & Steinhardt 1995). There is a potential fly in the ointment here, however, provided by the statistics of multiply lensed QSOs (see below), which seem to require $\Omega_\Lambda < 0.7$ at the present epoch. While a universe with $\Omega_0 + \Omega_\Lambda = 1$ and $\Omega_\Lambda > 0.7$ seems to be consistent with many of the other constraints we discuss in this book (Ostriker & Steinhardt 1995), it only just scrapes through the age problem: from equation (3.3), we find that the value of

† There is, in fact, another possibility which we discuss in Chapter 8.

$H_0 t_0$ exceeds that corresponding to an empty universe, $H_0 t_0 = 1$, only for $\Omega_\Lambda > 0.75$.

We went through the arguments in Chapter 2 as to why the introduction of a cosmological constant term is highly problematic from a physical point of view so we will not go through them again. From the point of view of the mathematical modeller, however, it is also unsatisfactory simply to introduce an extra free parameter into the model to explain one observation. Of course, we can fit the ages if we introduce a gratuitous extra parameter into the equations that has no observable effect except to solve the age problem. Unless some other corroborating evidence that could confirm the non-zero value of Λ is adduced, for example from galaxy number counts (see below), its introduction is *ad hoc* and to be avoided if at all possible, in keeping with Ockham's razor (Garrett & Coles 1993).

Finally, from a philosophical viewpoint, a (classical) cosmological constant goes against all that has been achieved in general relativity, where the geometry has been brought into the arena of physics: geometry is affected by matter, as well as matter being affected by the space-time geometry. A non-zero Λ affects all matter in the universe but is affected by nothing – violating the fundamental principle that anything that acts on matter should also be able to be acted on by matter. We choose to avoid this situation unless there is no other way out. Certainly we prefer solving the age problem by a low value of Ω_0 rather than a non-zero Λ, if it can be done.

On the other hand, if a cosmological constant term needs to be added in this way, the argument is essentially independent of the arguments emerging from inflation. If a Λ-term is to be introduced, one can imagine introducing it as a floating parameter, unconstrained by the requirement that $k = 0$. If one is prepared to permit this approach on pragmatic grounds, one can have a very old universe with negative spatial curvature, $\Omega_0 + \Omega_\Lambda < 1$, which satisfies the constraints we have discussed in this section. This choice of model may well turn out to be necessary if both H_0 and t_0 turn out to be at the upper end of the quoted limits, in such a way that $H_0 t_0 > 1$.

3.2 'Classical cosmology'

In the early days of observational cosmology, much emphasis was placed on the geometrical properties of expanding universe models as tools for estimating parameters of the cosmological models. Indeed, famous articles by Sandage (1968, 1970) reduced all cosmology to the task of determining two parameters: H_0, which we have discussed above, and q_0, the deceleration parameter. At a generic time t the deceleration parameter is defined by

$$q = -\frac{\ddot{a}a}{\dot{a}^2}; \tag{3.5}$$

as usual, the zero suffix means that q_0 is defined at the present time.

Matter-dominated models with vanishing Λ have

$$q_0 = \frac{\Omega_0}{2} \tag{3.6}$$

so the parameters q_0 and Ω_0 are essentially equivalent. If there is a cosmological constant contributing towards the spatial curvature, however, we have the general relation

$$q_0 = \frac{\Omega_0}{2} - \Omega_\Lambda. \tag{3.7}$$

In the case where $\Omega_\Lambda + \Omega_0 = 1$ ($\kappa = 0$) we have $q_0 < 0$ for $\Omega_0 < 2/3$.

The parameters H_0 and q_0 thus furnish a general description of the expansion of a cosmological model: these are Sandage's famous 'two numbers'. Their importance is demonstrated in standard cosmology textbooks (Weinberg 1972; Peebles 1993; Narlikar 1993; Coles & Lucchin 1995) which show how the various observational relationships, such as the angular diameter–redshift and apparent magnitude–redshift relations for standard sources, can be expressed in simple forms using these parameters and the Robertson–Walker metric, equation (1.1). In the standard Friedmann–Robertson–Walker models, the apparent flux density and angular size of a standard light source or standard rod depend in a relatively simple way on q_0 (Hoyle 1959; Sandage 1961, 1968, 1970, 1988; Weinberg 1972), but the relationships are more complex if the cosmological constant term is included (e.g. Charlton & Turner 1987). During the 1960s and early 1970s, a tremendous effort was made to determine the deceleration parameter q_0 from the magnitude-redshift diagram. For a while, the preferred value was $q_0 \simeq 1$ (Sandage 1968) but eventually the effort died away when it was realised that evolutionary effects dominated the observations; no adequate theory of galaxy evolution is available that could enable one to determine the true value of q_0 from the observations. To a large extent this is the state of play now, although the use of the angular size–redshift and, in particular, the magnitude-redshift relation for Type Ia supernovae have seen something of a renaissance of this method. We shall therefore discuss only the recent developments in the subsequent sections.

3.2.1 Standard candles

The fundamental property required here is the *luminosity distance* of a source, which, for models with $p = \Lambda = 0$, is given by

$$D_L(z) = \frac{c}{H_0 q_0^2}\left[q_0 z + (q_0 - 1)\left(\sqrt{2q_0 z + 1} - 1\right)\right]; \qquad (3.8)$$

this relationship is simply defined in terms of the intrinsic luminosity of the source L and the flux l received by an observer using the Euclidean relation

$$D_L = \left(\frac{L}{4\pi l}\right)^{1/2}. \qquad (3.9)$$

The dependence on cosmological parameters comes from the fact that light is spread out over a surface of (proper) area $4\pi r^2 a_0^2$, and the photons are both redshifted and time-dilated by the expansion as they journey from the source to the observer, so that $D_L = a_0^2 r / a(t)$. One then uses the Robertson–Walker metric with $\mathrm{d}s = 0$ to define the path of this light in space-time in terms of $\mathrm{d}t$:

$$\frac{c\mathrm{d}t}{a(t)} = \frac{\mathrm{d}r}{\sqrt{1 - \kappa r^2}}; \qquad (3.10)$$

the a-dependence can then be obtained from the Friedman equation (1.13), using the fact that

$$\frac{\kappa}{a_0^2} = (2q_0 - 1)H_0^2. \qquad (3.11)$$

Expanding (3.8) in a Taylor series yields

$$D_L = \frac{c}{H_0}\left[z + \frac{1}{2}(1 - q_0)z^2 + \ldots\right]. \qquad (3.12)$$

One usually seeks to exploit this dependence by plotting the so-called 'Hubble diagram' of apparent magnitude against redshift for objects of known intrinsic luminosity: this boils down to plotting $\log l$ against z, hence the dependence on D_L.

The problem with exploiting such relations to prove the value of q_0 directly is that one needs to have a standard 'candle': an object of known intrinsic luminosity. The dearth of classes of objects suitable for this task is, of course, one of the reasons why the Hubble constant is so poorly known locally. If it were not for recent developments based on one particular type of object – Type Ia supernovae – we would have been inclined to have omitted this section entirely. As it is now, we consider that these sources offer the most exciting prospects for classical cosmology within the next few years.

The homogeneity and extremely high luminosity of the peak magnitudes of Type Ia supernovae, along with physical arguments as to why they should be standard sources, have made these attractive objects for observational cosmologists in recent years (e.g. Branch & Tamman 1992), though the use of supernovae was discussed, for example, by Sandage (1961). The current progress stems from the realisation that these objects are not in fact identical, but form a family which can nevertheless be mapped onto a standard object by using independent observations. Correlations between peak magnitude and the shape of the light curve (Hamuy *et al.* 1995; Riess, Press & Kirshner 1995) or spectral features (Nugent *et al.* 1995) have reduced the systematic variations in peak brightness to about two-tenths of a magnitude. The great advantages of these objects are (i) because their behaviour depends only on the local physics, they are expected to be independent of environment and evolution and so are good candidates for standard candles, and (ii) that they are bright enough to be seen at quite high redshifts where the dependence on cosmological parameters (3.12) is appreciable: Perlmutter *et al.* (1996) discuss seven 'good' objects in the range $0.3 < z < 0.5$ displayed in Figure 3.1. The constraint imposed by these objects on a matter-dominated universe is that $\Omega_0 = 0.88^{+0.69}_{-0.60}$, while a flat model with a positive cosmological constant requires $\Omega_0 = 0.06^{+0.28}_{-0.34}$, both of these intervals corresponding to the 95 per cent confidence level, the main

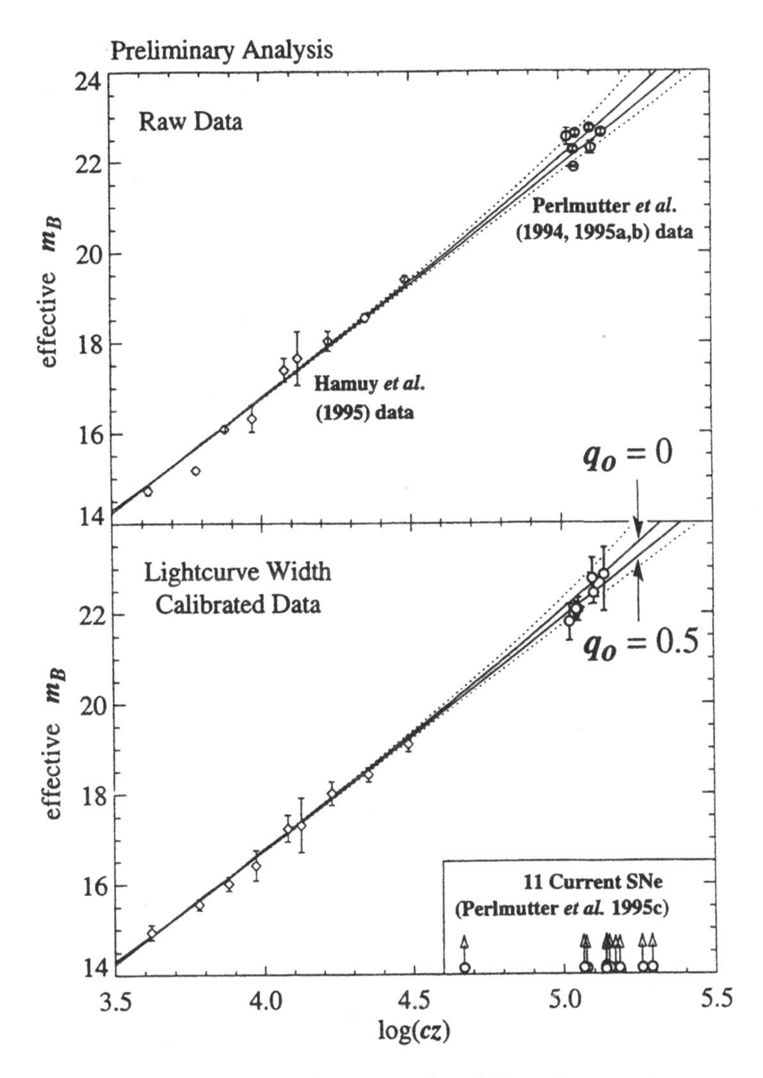

Fig. 3.1 Hubble diagrams for a sample of Type Ia supernovae reported by Perlmutter *et al.* (1996). These data place strong constraints on flat Λ-dominated models. Picture courtesy of Richard Ellis.

contribution to the uncertainty being the small sample size. The constraints on Λ-dominated models are thus severe indeed, but it is worth stressing that this is a relatively new method, and some systematic problems may well exist. Nevertheless, the rate at which supernovae are being found and the careful efforts being made to assess possible sources of error suggest that even stronger conclusions will be offered within a few years.

3.2.2 Angular sizes

The angle subtended by a standard metric 'rod' behaves in an interesting fashion as its distance from the observer is increased in standard cosmologies. It first decreases, as expected, then reaches a minimum after which it increases again (Sandage 1961). The position of the minimum depends upon q_0 (Ellis & Tivon 1985; Janis 1986). This somewhat paradoxical behaviour can be more easily understood by remembering that the light from very high redshift objects was emitted a long time ago when the proper distance to the object would have been much smaller than it is at the present epoch. Given appropriate dynamics, therefore, it is quite possible that distant objects appear larger than nearby ones with the same physical size.

For models with $\Lambda = 0$ the relationship between angular diameter θ and redshift z for objects moving with the Hubble expansion and with a fixed metric diameter d is simply

$$\theta = d\frac{(1+z)^2}{D_L(z)}, \tag{3.13}$$

where $D_L(z)$ is the *luminosity distance* given by equation (3.8).

Unfortunately, as with the standard candles, astronomers are generally not equipped with standard sources they are able to place at arbitrarily large distances. To try to use this method, one must select galaxies or other sources and hope that the intrinsic properties of the objects selected do not change with their distance from the observer. Because light travels with a finite speed, more distant objects emitted their light further in the past than nearby objects. Lacking an explicit theory of source evolution, one must assume the source properties do not vary with cosmological time. Since there is overwhelming evidence for strong evolution with

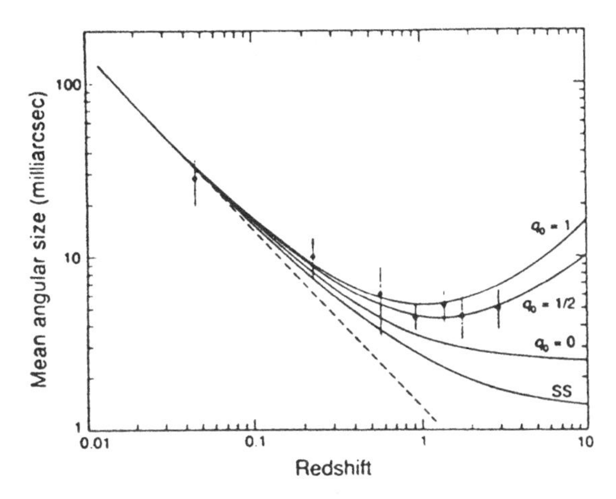

Fig. 3.2 Kellerman's data on angular sizes of compact radio sources. Reprinted, with permission, from Kellerman (1993). ©MacMillan Magazines Ltd (1993).

time in almost all classes of astronomical object, the prospects for using this method are highly limited.

A brave effort has been made recently by Kellerman (1993) to resurrect this technique by applying it to compact radio sources. These sources are much smaller than the extended radio sources discussed in previous studies, and Kellerman argues that one should therefore expect them to be less influenced by, for example, the evolution of the cosmological density. Kellerman indeed finds a minimum in the angular-size versus distance relationship, as shown in Figure 3.2, consistent with a high value of $q_0 \simeq 1/2$, and hence $\Omega_0 \simeq 1$. One should stress that this is essentially the only such analysis which has ever yielded a value of Ω_0 consistent with unity. Indeed a similar study by Gurvits (1994) gives $q_0 \simeq 0.16 \pm 0.71$, the most relevant point being the much larger errors. Actually a straightforward acceptance of the Kellerman data at its face value, using his own theoretical curves, suggests a best-fit value for Ω_0 *greater* than unity: between 1.5 and 2 (it was presumably theoretical reasons that prevailed against the author emphasising this implication of his results). If Ω_0 is actually unity, or close to it, some (unknown) systematic effect must be responsible for this apparent overestimate. But if the systematic error is this large, could it not be responsible for *all* the effect Kellerman observes?

However, a careful look at Kellerman's data (from a sample of only 82 sources) shows that he has not really detected a significant increase beyond the minimum. A model fit to his data in which the angular diameter decreases and then stays constant actually fits just as well as the $\Omega_0 = 1$ model. Since the evolution of radio sources – which are highly dynamic and active objects – is likely to be extremely complicated, it may be that Kellerman has simply discovered a different form of time evolution than that found for extended radio sources. Until it is clearly demonstrated that this is not the case (which, admittedly, would be very difficult to do), one would be unwise to place too much emphasis on this claim. Further criticisms of this claim can be found in Stepanas & Saha (1995) and Dabrowski *et al.* (1995).

We must therefore conclude that the evidence from the angular size data is not particularly compelling. Furthermore, we cannot see that there are in sight any 'standard metre sticks' that will be visible at high redshift and also will have well-understood evolutionary properties that could lead to a change in this situation. Thus we are not optimistic about this method, although it is possible that angular size estimates of clusters of sources, or measurments of angular separation of similar objects, could eventually give the statistical data needed for this test.

3.2.3 Number counts

An alternative approach is not to look at the properties of objects themselves but to try to account for the cumulative number of objects one sees in samples that probe larger and larger distances. A first application of this idea was by Hubble (1936); see also Sandage (1961). By making models for the evolution of the galaxy luminosity function one can predict how many sources one should see above an apparent magnitude limit and as a function of redshift. If one accounts for evolution of the intrinsic properties of the sources correctly, then any residual dependence on redshift is due to the volume of space encompassed by a given interval in redshift; this depends quite strongly on Ω_0. The considerable evolution seen in optical galaxies, even at moderately low redshifts, as well as the large K-corrections and uncertainties in the present-day luminosity function renders this type of analysis prone to all kinds of systematic uncertainties. One of the major problems here is that one does not have complete information about the redshift distribution of galaxies appearing in the counts. Without that information, one does not really know whether one is seeing intrinsically fainter galaxies relatively nearby, or relatively bright galaxies further away. This uncertainty makes any conclusions dependent upon the model of evolution assumed.

Controversies are rife in the history of this field. A famous recent application of this approach by Loh & Spillar (1986) yielded a value $\Omega_0 = 1^{+0.7}_{-0.5}$. This is, of course, consistent with unity but cannot be taken as compelling. A slightly later analysis of this data by Cowie (1987) showed how, with slightly different assumptions, one can reconcile the data with a much smaller value of Ω_0. Further criticisms of the Loh–Spillar analysis have been lodged by other authors (Koo 1988; Bahcall & Tremaine 1988; Caditz & Petrosian 1989). Such is the level and apparent complexity of the evolution in the stellar populations of galaxies over the relevant time-scale that we feel that it will be a long time before we understand what is going on well enough to even try to disentangle the cosmological and evolutionary aspects of these data. There has been significant progress, however, with number counts of faint galaxies, beginning in the late 1980s (Tyson & Seitzer 1988; Tyson 1988) and culminating with the famous 'deep field' image taken with the Hubble Space Telescope, which is shown in Figure 3.3. The 'state of the art' analysis of number counts (Metcalfe *et al.* 1996) is shown in Figure 3.4, which displays the very faint number counts from the HST in two wavelength bands, together with ground-based observations from other surveys. Although a variety of combinations of cosmological model and evolutionary scenarios seem to be allowed by the data, it seems fair to mention that at least the B counts appear to be more easily represented in low q_0 models. However, one should temper that claim by noting that the blue light is more likely to be dominated by transient massive stars and therefore might be expected to be more affected by evolution than other wavelengths. Although the HST and other number count data will no doubt provide *aficionados* of galaxy evolution plenty of constraints on their models (e.g. R. Ellis 1993), we have to conclude again that the direct evidence on Ω_0 is not compelling one

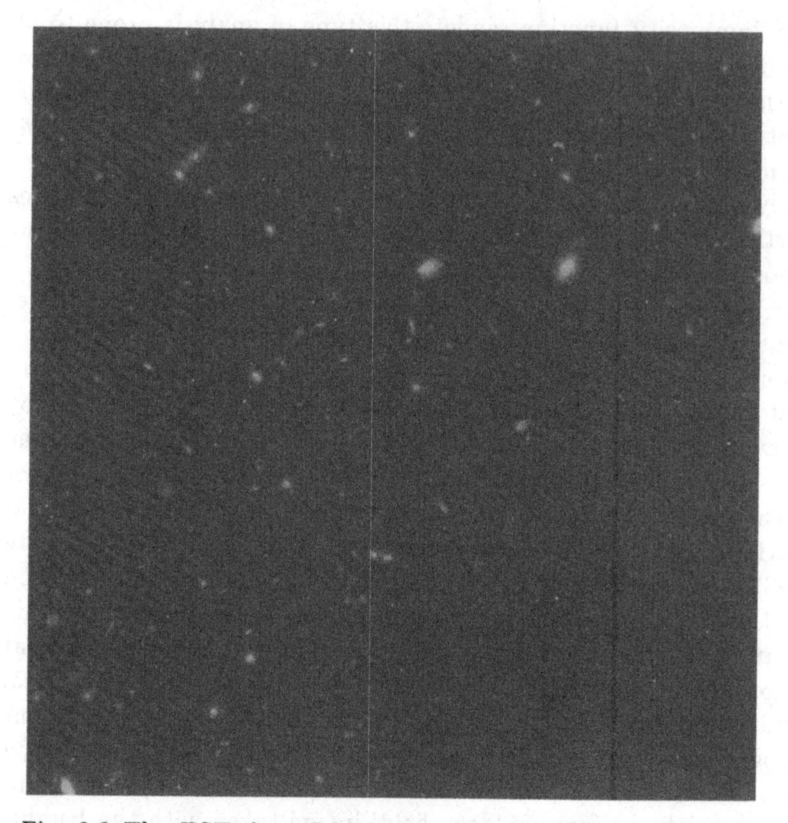

Fig. 3.3 The HST deep field image, showing images of galaxies down to limiting visual magnitude of about 28.5 in blue light. By extrapolating the local luminosity function of galaxies, one concludes that a large proportion of the galaxies at the faint limit have $z > 2$. Picture courtesy of Richard Ellis.

way or the other. The controversy surrounding the interpretation of the deep field results is compounded by uncertain estimates of the redshift distribution of these objects using photometric colours (Lanzetta, Yahil & Fernández-Soto; Metcalfe *et al.* 1996). The implications of these results for cosmological models are unlikely to be resolved unless and until there are major advances in the theory of galactic evolution. We have not discussed the properties of counts of quasars and radio-galaxies, which have been of great historical importance (Scott & Ryle 1961; Schmidt 1968), because

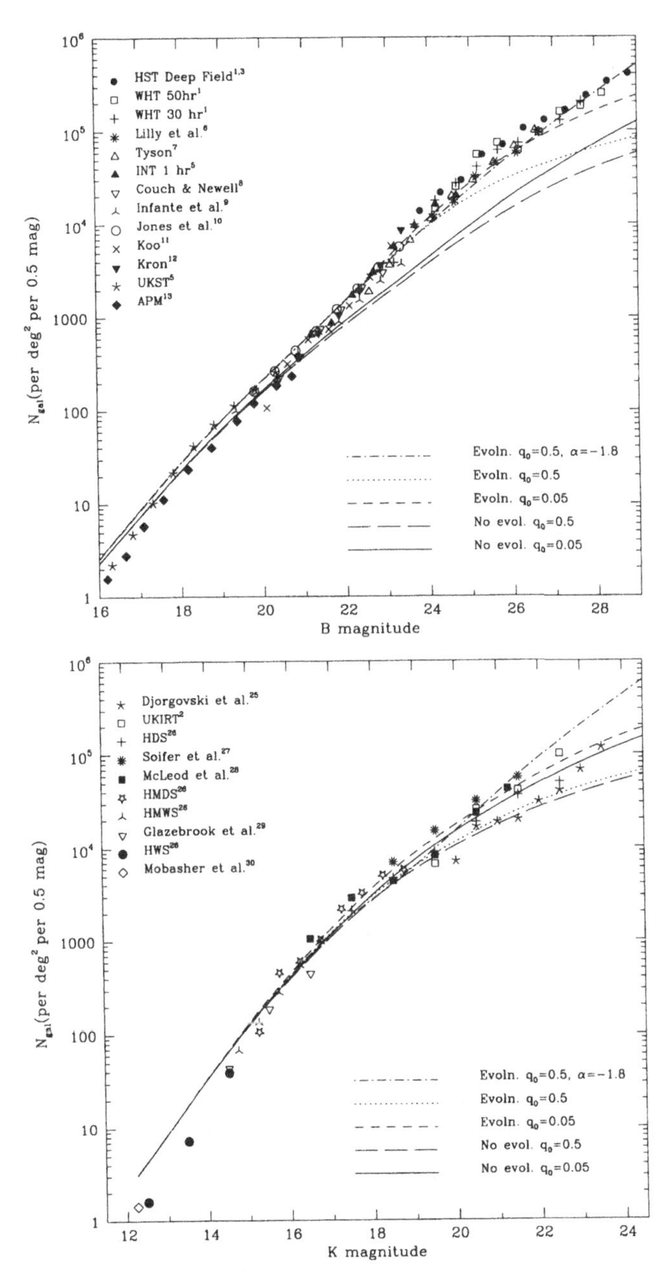

Fig. 3.4 Compilation of number-count data in the B (blue) and K (infrared) bands, from Metcalfe *et al.* (1996). Picture courtesy of Tom Shanks.

the evidence suggests that these too are also dominated by evolutionary effects and have consequently little to say directly about Ω_0.

3.3 Gravitational lensing

In principle the distortion of light paths owing to the gravitational field of astronomical bodies provides a way of probing the cosmological density of material in a much more direct way than through number-counts, angular sizes or other classical cosmological tests. In this chapter we shall restrict our discussion to what one might call *global* lensing phenomena; the use of lensing to probe the more local distribution of mass in invidual objects such as rich clusters and galactic haloes will be discussed, along with the other astrophysical evidence pertaining to these objects, in Chapter 5. In the context we are discussing here, there are basically two possible uses of lensing data.

3.3.1 Multiply imaged QSOs

The earliest known instance of gravitational lensing was the famous double quasar 0957+561, which upon close examination was found to be a single object which had been lensed by an intervening galaxy (Walsh, Carswell & Weymann 1979). As time has gone by, searches for similar such lensed systems have yielded a few candidates but no more than a handful. In the sample discussed by Kochanek (1993), for example, there are 680 quasars and only 5 lensed systems.

It has been known for some time that the predicted frequency of quasar lensing depends strongly on the volume out to a given redshift (Turner, Ostriker & Gott 1984; Turner 1990; Fukugita & Turner 1991) and that the number of lensed quasars observed can consequently yield important constraints on cosmological models. Compared to the Einstein–de Sitter model, both flat cosmologies with a cosmological constant and open low density ($\Omega_0 < 1$) models predict many more lensed systems. The effect is particularly strong for the flat Λ models: roughly ten times as many lenses are expected in such models than in the $\Omega_0 = 1$ case.

Of course, the number of lensed systems also depends on the number and mass of intervening objects in the volume out to the quasar, so any constraints to emerge are necessarily dependent upon assumptions about the evolution of the mass function of galaxies, or at least their massive haloes. Nevertheless, claims of robust constraints have been published (Kochanek 1993; Maoz & Rix 1993; Mao & Kochanek 1994) which constrain the contribution of a Λ term to the total density of a flat universe to $\Omega_\Lambda < 0.5$ at 90% confidence. Constraints on open, low-density models are much weaker: $\Omega_0 > 0.2$. Unless some significant error is present in the modelling procedure adopted in these studies, the QSO lensing statistics appear to rule out precisely those flat Λ-dominated models which have been held to solve the age problem and also allow flat spatial sections, although at a relatively low confidence level and at the expense of some model-dependence.

If our understanding of galaxy evolution improves dramatically it will be possible to refine these limits. New large-scale QSO surveys will also help improve the statistics of the lensed objects. At present all we can say is that they suggest that wise investors should not be buying shares in the $\Omega_0 + \Omega_\Lambda = 1$ models!

3.3.2 Microlensing

A slightly more subtle use of this general idea involves the phenomenon of microlensing. Even if the gravitational lens is not strong enough to form two or more distinct images of a background source, it may still amplify the brightness of the source to an observable extent. This idea has recently born fruit with convincing evidence for microlensing of stars in the LMC by massive objects in the galactic halo; we discuss this in the context of galactic halo dark matter in Chapter 5.

A more ambitious claim by Hawkins (1993) to have observed microlensing on a cosmological scale by looking at quasar variability is much less convincing. To infer microlensing from quasar light curves requires one to exclude the possibility that the variability seen in the light curves be intrinsic to the quasar. While it is true that one might naively expect the time-scale of intrinsic variability to increase with QSO redshift as a result of the cosmological time dilation and this is not seen in the data, this is but one of many effects which could influence the time-scale. For example, the density of cosmological material surrounding a QSO increases by a factor of eight between $z = 1$ and $z = 3$. Alexander (1995) gives arguments that suggest observational selection effects may remove the expected correlation and replace it with the inverse effect that is actually observed. It is not inconceivable therefore that a change in fuelling efficiency could change the time-scale of variability in the opposite direction to the time dilation effect. In any case, the classic signature of microlensing is that the variability be achromatic: even this is not known about the variability seen by Hawkins. QSOs, and active galaxies in general, exhibit variability on a wide range of time-scales in all wavelength regions from the infrared to X-rays. If the lensing interpretation is correct then one should be able to identify the same time-scale of variability at all possible observational wavelengths. An independent analysis of QSO variability by Dalcanton *et al.* (1994) has also placed Hawkins' claim in doubt, so we take the evidence that extragalactic microlensing has been detected to be rather tenuous.

3.4 Summary

We end this chapter by summing up the strength of the evidence from the various considerations we have discussed

• **Ages.** This is probably the most fundamental problem facing modern cosmology. It is no exaggeration to say that the subject will be poised on the brink of a crisis if H_0 and t_0 both turn out to be at the upper limit of the allowed error ranges. Although there are still substantial uncertainties, and history warns us against taking estimates of H_0 as gospel truth, we do think that there is a shift in the weight of evidence away from a relatively young universe with $\Omega_0 \simeq 1$ and towards an older universe with a relatively low matter-density (e.g. Ellis *et al.* 1996).

• **Classical cosmology.** Being conservative, we have concluded that the data coming from number count analyses and angular diameters are probably not sufficiently reliable or model-independent to draw any firm conclusions from them. Recent work on the use of Type Ia supernovae as standard candles on the other hand looks very good, and these objects offer a definite prospect for the future. Present constraints from this technique suggest that the allowed range of q_0 can be narrowed substantially in the near future.

• **Gravitational lensing.** Global lensing arguments only really constrain Λ at the present, but may offer the way to rule out the 'inflationary' models with $\kappa = 0$ and $\Omega_0 < 1$. It is possible that such arguments may rule out values of Ω_Λ large enough to reconcile stellar evolution ages with H_0 and also with $\kappa = 0$, although uncertainties in modelling the evolution of galactic haloes need to be resolved before this result can be taken as definitive.

As a final comment we should point out that (a) we have not considered the complex selection effects and statistical analyses that dominate understanding of these observational relations, and (b) many of the arguments we have considered in this chapter depend on the assumption of large-scale homogeneity. Inhomogeneities, which we know exist, may well affect the interpretation of local values of the Hubble parameter as well as observational relations. We return to the latter issue and related questions in Chapter 8.

4
Element abundances

One of the great successes of the hot-big-bang model is the agreement between the observed abundances of light elements and the predictions of nucleosynthesis calculations in the primordial fireball. However, this agreement can only be made quantitative for certain values of physical parameters, particularly the number of light neutrino types, neutron half-life and cosmological entropy-per-baryon. Since the temperature of the cosmic microwave background radiation is now so strongly constrained, the latter dependence translates fairly directly into a dependence of the relative abundances of light nuclei upon the contribution of baryonic material to Ω_0. It is this constraint, the way it arises and its implications that we shall discuss in this chapter. For more extensive discussions of both theoretical and observational aspects of this subject, see the technical review articles of Boesgaard & Steigman (1985), Bernstein *et al.* (1988), Walker *et al.* (1991) and Smith *et al.* (1993).

4.1 Theory of nucleosynthesis

4.1.1 Prelude

We begin a brief description of the standard theory of cosmological nucleosynthesis in the framework of the big-bang model with some definitions and orders of magnitude. The *abundance by mass* of a certain type of nucleus is the ratio of the mass contained in such nuclei to the total mass of baryonic matter contained in a suitably large volume. As we shall explain, the abundance of ^{4}He, usually indicated with the symbol Y, has a value $Y \simeq 0.25$, or about 6% of all nuclei, as determined by various observations (of diverse phenomena such as stellar spectra, cosmic rays, globular clusters and solar prominences). Other abundances are usually expressed relative to the abundance of H, e.g. D/H. The abundance of ^{3}He is of the order of 10^{-5}, similar (at least to an order of magnitude) to that of deuterium D (i.e. ^{2}H), while the abundance of lithium (^{7}Li) is of the order of 10^{-10}. Notice that these abundances span at least nine orders of magnitude. We shall see that the predicted abundances vary systematically with the baryon-to-photon ratio, η, and hence with the fraction of the critical density in the form of baryons, Ω_b. These two quantities are related through the expression

$$\Omega_b h^2 = 0.004\eta_{10}, \tag{4.1}$$

where we have used our usual notation for the Hubble parameter h and $\eta = \eta_{10} \times 10^{-10}$. The accuracy with which the present temperature of the microwave background is known limits the uncertainty in the numerical coefficient on the right-hand side of equation (4.1) to a few per cent. Other dependences we shall discuss below relate to the neutron lifetime τ_n (Byrne *et al.* 1990) and the number of light neutrino types N_ν (Dydak 1991), both of which are now strongly constrained by laboratory experiments.

In the standard cosmological model, the nucleosynthesis of the light elements (which we take to mean elements with nuclei no more massive than ^{7}Li) begins at the start of the radiative era when the temperature $T \sim 10^9$ K. Although nuclear fusion begins only a matter of seconds after the big bang, it is important to stress that it does represent well-known physics, in the sense that relevant time-scales are long and energies well below those that have been probed in laboratory experiments. Moreover, the separation between nucleons is considerably larger than the nuclear length scale so that collective and many-body effects are expected to be negligible. As we shall see, the reaction network is also relatively simple as it involves only nuclei with mass numbers below that of ^{7}Li.

Nucleosynthesis of the elements also occurs in stellar interiors during the course of stellar evolution. Stellar processes generally involve a destruction of D more quickly than it is produced, because of the very large cross-section for photodissociation reactions of the form

$$\mathrm{D} + \gamma \rightleftharpoons p + n. \tag{4.2}$$

Nuclei heavier than ^{7}Li are essentially only made in stars. In fact there are no stable nuclei with atomic weight 5 or 8 so it is difficult to construct elements heavier than helium by means of $p + \alpha$ and $\alpha + \alpha$ collisions (α represents a ^{4}He nucleus). In stars, however, $\alpha + \alpha$ collisions do produce small quantities of unstable ^{8}Be, from which one can make ^{12}C by ^{8}Be$+\alpha$ collisions; a chain of synthesis reactions can therefore develop leading to heavier elements. In the cosmological context, at the temperature of 10^9 K characteristic of the onset of nucleosynthesis, the density of the universe is too low to permit the synthesis of significant amounts of ^{12}C from ^{8}Be$+\alpha$ collisions. It turns out therefore that the elements heavier than ^{4}He are made mostly in stellar interiors. On the other hand, the percentage of helium observed is too high to be explained by the usual predictions of stellar evolution. For example, if our galaxy maintained a constant luminosity for the order of 10^{10} years, the total energy radiated would correspond to the fusion† of one per cent of the original nucleons, in contrast to the six per cent which is observed.

It is interesting to note that the difficulty in explaining the nucleosynthesis of *observed levels* of helium by stellar processes alone was recognised as early as the 1940s (Gamow 1946; Alpher & Herman 1948; Alpher, Bethe & Gamow 1948). Difficulties with this model, in particular an excessive production of helium, persuaded these authors to consider the idea that there might have been a significant radiation background at the epoch of nucleosynthesis; they estimated that this background should have a present temperature of around 5 K, not far from the value it is now known to have ($T \simeq 2.73$ K), although some 15 years were to intervene before this background was discovered; see also Hoyle & Tayler (1964). For this reason one can safely say that the satisfactory calculations of primordial element abundances which emerge from the theory represent, along with the existence of the cosmic radiation background, one of the central pillars upon which the big-bang model is based.

† It is possible that a significant part of the nucleosynthesis of helium might take place in an initial highly luminous phase of galaxy formation or, perhaps, in primordial 'stars' of very high mass (Bond, Carr & Arnett 1983). Such models are, however, subject to strong observational constraints from, for example, the infrared background; see Carr (1994) for a review.

4.1.2 Big-bang nucleosynthesis

Although the physics of primordial nucleosynthesis is relatively simple, a number of hypotheses need to be made in order that a well-defined model be constructed. The most important of these assumptions are:

(a) the universe has passed through a hot phase with $T \geq 10^{12}$K, during which its components were in thermal equilibrium;
(b) general relativity and other 'known' laws of particle physics apply at this time;
(c) the universe is homogeneous and isotropic at the time of nucleosynthesis;
(d) the number of neutrino types is not high (in fact we shall asssume $N_\nu \simeq 3$);
(e) the neutrinos have a negligible degeneracy parameter;
(f) the universe is not composed in such a way that some regions contain matter and others antimatter;
(g) there is no appreciable magnetic field at the epoch of nucleosynthesis;
(h) the density of any exotic particles (photinos, gravitinos, etc.) at relevant times is negligible compared to the density of the photons.

These hypotheses agree pretty well with such facts as we know. Hypothesis (c) is made because at the moment of nucleosynthesis the mass of baryons contained within the cosmological horizon is very small (i.e. less than $10^3\ M_\odot$), while the light element abundances one measures seem to be the same over scales of the order of tens of Mpc; hypotheses (d) and (h) are necessary because an increase in the density of the universe at the epoch of nucleosynthesis would lead, as we shall see, to an excessive production of helium; hypothesis (f) is made because the gamma rays which would be produced at the edges where such regions touch would result in extensive photodissociation of the D, and a consequent decrease in the production of ^{4}He. We shall discuss non-standard nucleosynthesis models, in which one or more of these assumptions can be varied, in §4.3.

The assumption of thermal equilibrium (a) allows one to invoke standard statistical mechanics arguments to give the number-densities of relevant particles in terms of Boltzmann distributions:

$$n_i \simeq g_i \frac{(m_i k_B T/2\pi)^{3/2}}{\hbar^3} \exp\left(\frac{\mu_i - m_i c^2}{k_B T}\right), \tag{4.3}$$

with i being a label standing for for the particle species in question, μ_i being its chemical potential and g_i being the relevant number of spin states (e.g. Bernstein 1988). At the beginning of the radiative era, the most important consideration is the relative abundance of neutrons and protons, held in equilibrium by the weak nuclear interactions

$$n + \nu_e \rightleftharpoons p + e^-, \quad n + e^+ \rightleftharpoons p + \bar{\nu}_e, \tag{4.4}$$

which occur on a characteristic time-scale $\tau_{\rm coll}$ that is much smaller than the expansion time-scale $\tau_H = (a/\dot{a})$ until the temperature drops to $T_\nu \simeq 10^{10}$ K, i.e. when the neutrinos decouple from matter. Because of the assumption of thermal equilibrium and non-degeneracy of the neutrinos, the ratio of the number densities of neutrons and protons is given by

$$\frac{n_n}{n_p} \simeq \exp\left(-\frac{Q}{k_B T}\right) = \exp\left(-\frac{1.5 \times 10^{10}\ \mathrm{K}}{T}\right), \tag{4.5}$$

where Q is the 'binding energy' of the neutron: $Q = \Delta m c^2$, with Δm equal to the difference in rest masses of the neutron and proton. At T_ν this ratio is

$$X_n(T_\nu) = \frac{n}{n+p} \simeq \frac{n}{n_{tot}} \simeq 0.17; \tag{4.6}$$

accurate calculations show that the ratio X_n remains roughly equal to this equilibrium value until $T_n \simeq 1.3 \times 10^9$K, after which the neutrons can only transform into protons via β-decay:

$$n \rightarrow p + e^- + \bar{\nu}_e, \tag{4.7}$$

which has a mean lifetime τ_n of the order of 900 s. Below T_n (i.e. for $t > t_n$) the ratio X_n then varies according to the exponential law of radioactive decay:

$$X_n(t) \equiv X_n(0) \exp\left(-\frac{t - t_n}{\tau_n}\right), \tag{4.8}$$

for $t - t_n \simeq t < \tau_n$. The time-scales for change of temperature, determined by the Friedmann equation (1.13) with $\rho = \rho_r = aT^4$ (the matter density and curvature terms being negligible at this time), are such that at the time nucleosynthesis begins, X_n has altered only a little from its initial value given by equation (4.6).

4.1.3 Helium 4

When the temperature is of the order of T_n, the relevant components of the universe are photons, protons and neutrons in the ratio discussed above, and small amounts of heavier particles (besides the neutrinos which have already decoupled). To build nuclei with atomic weight $A \geq 3$ one needs first to generate a significant amount of deuterium. The amount created is governed by the equation

$$n + p \rightleftharpoons \mathrm{D} + \gamma : \tag{4.9}$$

this reaction has a characteristic time-scale $\tau_{\rm coll} \ll \tau_H$ in the period under consideration. Because of the equilibrium conditions, the chemical potentials for the equation (4.9) are related by

$$\mu_n + \mu_p = \mu_{\rm D}. \tag{4.10}$$

It is relatively straightforward then to calculate the quantities $X_p = p/n_{\rm tot} \simeq 1 - X_n$ and $X_{\rm D} = \mathrm{D}/n_{\rm tot}$, which depend on the temperature $\Omega_{\rm b}$ and B_d, the binding energy of deuterium:

$$B_d = (m_n + m_p - m_{\rm D})c^2 \simeq 2.225 \text{ MeV} \simeq 2.5 \times 10^{10} \text{ K}. \tag{4.11}$$

The resulting X_d depends only weakly on Ωh^2. For $T \geq 10^{10}$K the value of $X_{\rm D}$ is negligible: all the nucleons are still free because the high energy of the ambient photons favours the photodissociation reaction. The fact that nucleosynthesis cannot proceed until $X_{\rm D}$ grows sufficiently large is usually called the *deuterium bottleneck* and is an important influence on the eventual helium abundance. The value of $X_{\rm D}$ is no longer negligible when the temperature falls to some temperature $T^* \simeq 10^9$ K, defining the onset of nucleosynthesis at $t = t^*$. At lower temperatures still all the neutrons might be expected to be captured to form deuterium. This deuterium does not appear, however, because reactions of the form

$$\mathrm{D} + \mathrm{D} \rightarrow {}^3\mathrm{He} + n, \qquad {}^3\mathrm{He} + \mathrm{D} \rightarrow {}^4\mathrm{He} + p, \tag{4.12}$$

which have a large cross-section and are therefore very rapid, mop up any free neutrons into ^{4}He. Thus the abundance of helium that forms is

$$Y \simeq Y(T^*) = 2X_n(T^*) = 2X_n(T_n)\exp\left(-\frac{t^* - t_n}{\tau_n}\right) \simeq 0.25, \tag{4.13}$$

using plausible values for t^* and t_n. The result (4.13) is in reasonable accord with that given by observations; see §4.2. In equation (4.13), the factor 2 takes account of the fact that, after helium synthesis, there are practically only free protons and helium nuclei, so that

$$Y = \frac{m_{\rm He}}{m_{\rm tot}} = 4\frac{n_{\rm He}}{n_{\rm tot}} \simeq 4 \times \frac{1}{2}\frac{n_n}{n_{\rm tot}} = 2X_n. \qquad (4.14)$$

The value of Y obtained is thus roughly independent of η or $\Omega_{\rm b}$. This is essentially for two reasons: (i) the value of X_n before nucleosynthesis does not depend on Ω because it is determined by weak interactions between nucleons and leptons and not by strong interactions between nucleons; (ii) the start of nucleosynthesis is determined by the temperature rather than the number-density of the nucleons.

These calculations show why the ^{4}He abundance is potentially so important – it is, to a good approximation, independent of η (or $\Omega_{\rm b}$) as well as the 'ancillary' parameters τ_n and N_ν. It is therefore a useful consistency check on the whole scheme. More detailed calculations show that the dependence on these parameters is of the form

$$Y \simeq 0.225 + 0.012(N_\nu - 3) + \log_e \eta_{10} + 0.0097(\tau_n - 14.9), \quad (4.15)$$

with τ_n measured in minutes (Walker *et al.* 1991).

4.1.4 Other light nuclei

As far as the abundances of other light elements are concerned one needs to perform a detailed numerical integration of all the rate equations describing the reaction network involved in building up nuclei heavier than ^{4}He. The numerical calculations shown in Figure 4.1 demonstrate that the abundance of deuterium is much more steeply dependent upon η:

$$\mathrm{D/H} = 5 \times 10^{-4}\eta_{10}^{-5/3}; \qquad (4.16)$$

the abundance of ^{3}He behaves in a similar manner. The graph also shows the combined abundances of D and ^{3}He for reasons we shall discuss in a little more detail in the next section – essentially because stellar processes can destroy D but tend to compensate by producing ^{3}He, so the combined abundance of these two gives an upper limit on the *primordial* abundance of D. The basic effect

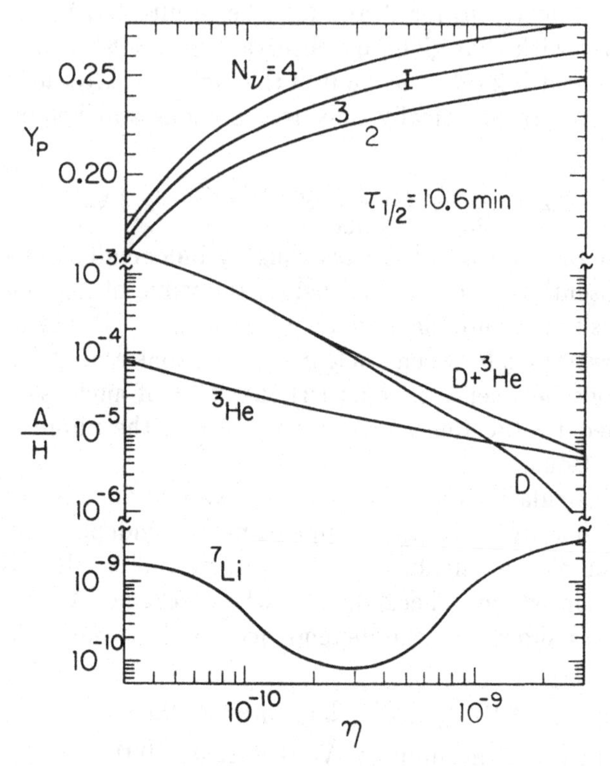

Fig. 4.1 Numerical calculations of light element abundances as a function of η. The helium abundance is also shown as a function of N_ν. From Kolb & Turner (1990). Copyright Addison-Wesley Publishing Company, Inc. Reproduced by permission of the publisher.

one can see is that, since the abundance of ^{4}He increases slowly with η (because nucleosynthesis starts slightly earlier and burning into ^{4}He is more complete), the abundances of the 'incomplete' products D and ^{3}He decrease in compensation. The abundance of ^{7}Li is rather more complicated because of two possible formation mechanisms: direct formation via fusion of ^{4}He and ^{3}H dominates at low η, while electron capture by ^{7}Be dominates at high η. In between, the 'dip' is caused by a destruction reaction involving proton capture and decay into two ^{4}He nuclei that is efficient at intermediate values of η.

4.2 The evidence

So how do these computations compare with observations? At the outset we should stress that relevant observational data in this field are difficult to obtain. For example, the expected abundance of ^{4}He is relatively large, but its dependence on cosmological parameters is not strong. Precise measurements are therefore required to test the theory. For the other elements shown in Figure 4.1, the parameter dependence is strong and is dominated by the dependence on η, but the expected abundances, as we have shown, are tiny. Moreover, any material we can observe has been at least partly processed through stars. Burning of H into ^{4}He is the main source of energy for stars. Deuterium can be very easily destroyed in stars (but cannot be made there). The other isotopes ^{3}He and ^{7}Li can be both created and destroyed in stars. The processing of material by stars is called *astration* and it means that uncertain corrections have to be introduced to derive 'primordial' abundances from present-day observations. One should also mention that *fractionation* (either physical or chemical in origin) may mean that the abundances in one part of an astronomical object may not be typical of the object as a whole. For instance, such effects are known to be important in determining the abundance of deuterium in the Earth's oceans.

Despite these considerable difficulties, there is a large industry involved in comparing observed abundances with these theoretical predictions. Relevant data can be obtained from stellar atmospheres, interstellar emission and absorption lines (and intergalactic ones), planetary atmospheres, meteorites and terrestrial measurements. Abundances of elements other than ^{4}He determined by these different methods differ by a factor of five or more, presumably because of astration and/or fractionation. The observational situation is therefore somewhat difficult to interpret and this issue would require a whole book to be devoted to it for it to be described fairly. What we shall do here, therefore, is simply to mention some of the most recent relevant observations and their uncertainties.

4.2.1 Helium 4

It is relatively well established that the abundance of ^{4}He is everywhere close to 25% and this in itself is good evidence that the basic model is correct. To get the primordial helium abundance more accurately than this rough figure, it is necessary to correct for the processing of hydrogen into helium in stars. This is generally done by taking account of the fact that stars with higher metallicity have a slightly higher ^{4}He abundance, and extrapolating to zero metallicity; metals are assumed to be a by-product of the fusion of hydrogen into helium. One therefore generally requires an index of metallicity in the form of either O/H or N/H determinations. Good data on these abundances have been obtained for around 50 extragalactic HII regions (Pagel *et al.* 1992; Skillman *et al.* 1993; Izatov *et al.* 1994). Olive & Steigman (1995) and Olive & Scully (1995), for example, have found on the basis of these data that there is evidence for a linear correlation of Y with O/H and N/H; the intercept of this relation yields

$$Y_p = 0.234 \pm 0.003 \pm 0.005. \tag{4.17}$$

The first error is purely statistical and the second is an estimate of the systematic uncertainty in the abundance determinations.

4.2.2 Deuterium

The abundace of deuterium has been the subject of intense investigation in recent months. Prior to this period, deuterium abundance information was based on interstellar medium (ISM) observations and Solar System data. From the ISM, one gets

$$\mathrm{D/H} \simeq 1.60 \times 10^{-5} \tag{4.18}$$

with an uncertainty of about 10% (Linsky *et al.* 1993, 1995). This value may or may not be close to universal, as it is possible that the abundances in the ISM are inhomogeneous. Solar System investigations based on properties of meteoritic rock involve a more circuitous route through ^{3}He (which one assumes was efficiently burned into D during the pre-main-sequence phase of the Sun). This argument leads to a value of

$$(\mathrm{D/H})_\odot \simeq 2.6 \times 10^{-5} \tag{4.19}$$

with an uncertainty of at nearly 100% (Scully *et al.* 1996).

More recently, the rough consensus between these two estimates has been shaken by (still controversial) claims of detections of deuterium absorption in the spectra of high-redshift quasars. The occurrence of gas at high redshift and in systems of low metallicity suggests that one might well expect to see a light element abundance close to the primordial value. The first such observations yielded much higher values than (4.18) and (4.19) by about a factor of 10 (Carswell *et al.* 1994; Songaila *et al.* 1994), i.e.

$$(\mathrm{D/H}) \simeq 2 \times 10^{-4}; \tag{4.20}$$

other measurements seem to confirm these high values (Rugers & Hogan 1996a,b; Carswell *et al.* 1996; Wampler *et al.* 1996). On the other hand, significantly lower deuterium abundances have been found by other workers in similar systems (Tytler *et al.* 1996; Burles & Tytler 1996). This raises the suspicion that the high inferred deuterium abundances may be a mistake, perhaps from a misidentified absorption feature (e.g. Steigman 1984). On the other hand, one does expect deuterium to be destroyed by astration and, on these grounds, one is tempted to identify the higher values of D/H with the primordial value. The situation remains unclear.

4.2.3 Helium 3

There are various ways in which the primordial ^{3}He abundance can be estimated. For a start, the Solar System deuterium estimate entails an estimate of the ^{3}He abundance which generally comes out around 1.5×10^{-5}. ISM observations and galactic HII regions yield values with a wide dispersion:

$$(^3\mathrm{He/H}) \simeq 2.5 \times 10^{-5}; \tag{4.21}$$

the spread is around a factor of 2 either side of this value.

The primordial ^{3}He, however, is modified by the competition between stellar production and destruction processes, and a detailed evolution model is required to relate the observed abundances, themselves highly uncertain, with their inferred primordial values. As we mentioned above, one may be helped in this task by using the combined abundance of D and ^{3}He (e.g. Steigman & Tosi 1994). The simplest way to use this data employs the argument that when deuterium is processed into stars it is basically turned into ^{3}He, which can be processed further, but which burns at a higher temperature. Stars of different masses therefore differ in their net conversion between these two species. But since *all* stars do destroy deuterium to some extent and at least *some* ^{3}He survives stellar processing, the primordial combination of D $+^3$ He might well be expected to be bounded above by the observed value. Attempts to go further introduce further model-dependent parameters and corresponding uncertainties into the analysis. For reference, a rough figure for the combined abundance is

$$(\mathrm{D} +^3 \mathrm{He})/\mathrm{H} \simeq 4.1 \times 10^{-5}, \tag{4.22}$$

with an uncertainty of about 50%.

4.2.4 Lithium 7

In old hot stars (Population II), the lithium abundance is found to be nearly uniform (Molaro *et al.* 1995; Spite *et al.* 1996). Indeed there appears to be little variation from star to star in a sample of 100 halo stars, over and above that expected from the statistical errors in the abundance determinations. The problem with the interpretation of such data, however, is in the fact that astrophysical processes can both create and destroy lithium. Up to about half the primordial ^{7}Li abundance may have been destroyed in stellar processes, while it is estimated that up to 30% of the observed abundance might have been produced by cosmic ray collisions. The resulting best guess for the primordial abundance is

$$\mathrm{Li}/\mathrm{H} \simeq 1.6 \times 10^{-10}, \tag{4.23}$$

but the uncertainty, dominated by unknown parameters of the model used to process the primordial abundance, is at least 50% and is itself highly uncertain (Walker *et al.* 1993; Olive & Schramm 1993; Steigman *et al.* 1993).

4.2.5 Observations vs. theory

We have tried to be realistic about the uncertainties in both the observations and the extrapolation of those observations back to the primordial abundances. Going into the detailed models of galactic chemical evolution that are required to handle D, ^{3}He and ^{7}Li opens up a rather large can of model-dependent worms, so we shall simply sketch out the general consensus about what these results mean for η and Ω_b.

Setting aside the rather controversial claims of a high primordial deuterium abundance for the moment, one sees that the estimates of the primordial values of the relative abundances of deuterium (D), ^{3}He, ^{4}He and ^{7}Li all appear to be in accord with nucleosynthesis predictions, but only if the density parameter in baryonic material is in the range

$$0.01 < \Omega_b h^2 < 0.015 \tag{4.24}$$

(e.g. Walker *et al.* 1991; Smith *et al.* 1993). This roughly corresponds to $3 \leq \eta_{10} \leq 4$. A baryon density higher than this would produce too much ^{7}Li, while a lower value would produce too much deuterium and ^{3}He. Copi *et al.* (1995a,b) suggest a somewhat wider range of allowed systematic errors, leading to $2 \leq \eta_{10} \leq 6.5$, which translates into

$$0.005 < \Omega_b h^2 < 0.026. \tag{4.25}$$

The dependence of ^{4}He is so weak that it can really only be used as a consistency check on the scheme. There is potentially a problem here because, for the value of η_{10} required to match the above abundances, the predicted value of Y_p for the range (4.24) is discrepant at the 2σ (statistical errors only) level with respect to the inferred value. Although the statistical errors on Y_p are relatively small, we do not feel that this discrepancy constitutes much of a crisis because only a small systematic artefact would be required to bring the two values into agreement. We shall adopt the second, looser limit (4.25) here.

If we are prepared to take the high deuterium abundance as the gospel truth, however, the situation changes rather. Rugers & Hogan (1996a,b) argue that their best estimate for the primordial D/H abundance, allowing only modest destruction of deuterium, because of the very low metallicities of the systems involved, is

$$\mathrm{D/H} \simeq 10.9 \pm 0.4 \times 10^{-4}. \tag{4.26}$$

This leads to $\eta_{10} = 1.7 \pm 0.2$, and consequently

$$0.0054 < \Omega_b h^2 < 0.007. \tag{4.27}$$

This is then more consistent with the inferred primordial value of Y_p, but apparently discrepant with ^{3}He. Since the interpretation of the data on this latter value is so much more questionable, however, one should not be worrying too much about a possible deuterium crisis.

4.3 Non-standard nucleosynthesis

We have seen that standard nucleosynthesis seems to account reasonably well for the observed light element abundances and also places strong constraints on the allowed range of the density parameter. In our usual role as devil's advocates, we now ask ourselves the question to what extent these results rule out alternative models for nucleosynthesis, and what constraints can we place on models which violate conditions (a) to (h) of the previous section. We shall make some comments on this question by describing some attempts that have been made to vary the conditions pertaining to the standard model. The following discussion is not intended to be exhaustive.

First, one could seek to change the expansion rate τ_H at the start of nucleosynthesis (determined by (1.13) for $\rho = M/a^3$, $\kappa_0/a^2 \simeq 0$). A decrease of τ_H (i.e. a faster expansion rate) can be obtained if the universe contains other types of particles in equilibrium at the epoch under consideration, leading to an effective added contribution to ρ in (1.13) at that time. These could include new types of neutrino, or supersymmetric particles like photinos and gravitinos, an additional source of (perhaps anisotropic) pressure such as a magnetic field. More generally, a change of expansion rate would occur in an intrinsically anistropic universe (Thorne 1967; Barrow 1976). A small reduction of τ_H reduces the time available for the neutrons to decay into protons, so that the value of X_n tends to move towards its initial (high-temperature) value of $X_n \simeq 0.5$; the reduction of τ_H does not, however, influence the time of onset of nucleosynthesis to any great extent so that this still occurs at $T \simeq 10^9$ K. The net result (Peebles 1965) is an increase in the amount of ^{4}He produced. As we have mentioned above, these results have for a long time led cosmologists to rule out the possibility than N_ν might be larger than 4 or 5. Now we know that $N_\nu \simeq 3$ from particle experiments (Dydak 1991); nucleosynthesis still rules out the existence of any other relativistic particle species at the appropriate epoch, and places stringent constraints on the possible presence of primordial magnetic fields.

A large reduction in τ_H, however, tends to reduce the abundance of helium: the reactions (4.12) have too little time to produce significant helium for the density of the universe falls rapidly. A decrease in the expansion rate allows a larger number of neutrons to decay into protons so that the ratio $X_n(T^*)$ becomes smaller. Since basically all the neutrons end up in helium, the production of this element is decreased.

Even the assumption that the dynamics of the universe are described by general relativity is open to question. Inflation theorists often incorporate extended gravity theories in their models. If these are correct, then it is is possible that there might remain significant effects of the altered gravity law even as late as the epoch of nucleosynthesis. Such models have been investigated by Serna & Alimi (1996a,b) and they do indeed seem to offer significant possibilities to alter the link between η_{10} and the element abundances which is inherent in the standard model.

Another modification one can consider concerns the hypothesis that the neutrinos are not degenerate (e.g. Coles & Lucchin 1995). Apart from the apparent lack of 'naturalness' of this assumption from the point of view of particle physics, the problem here may be one of fine-tuning. If the chemical potential of the electron neutrinos is too large and positive then the neutron–proton ratio becomes very small and one consequently makes hardly any helium. If, on the other hand, the degeneracy parameter is too large in a negative sense, the high number of neutrons and the consequent low number of protons prevents the formation of deuterium and therefore helium. Deuterium would be formed much later when the expansion of the universe had diluted the $\bar{\nu}_e$ and some neutrons could have decayed into protons. But at this point the density would be too low to permit significant nucleosynthesis, unless $\Omega_b \geq 1$. In between, one can imagine a situation where one can have $X_n \simeq 0.5$ at the moment of nucleosynthesis, so that all the neutrons end up in helium, with the result that essentially all the baryonic matter in the universe would be in the form of helium. In the case of nucleosynthesis where the neutrinos or antineutrinos are degenerate there is another complication in the theory: the total density of neutrinos and antineutrinos would be greater than in the non-degenerate case. This gives rise to a decrease in the characteristic time for the expansion τ_H, with the corresponding consequences for nucleosynthesis. One can therefore conclude that the problems connected with a significant neutrino degeneracy are large, and one might be tempted to reject them on the pragmatic theorist's grounds that such models are also much more complicated than the standard model.

There is also the possibility that the decay of an unstable particle after the usual epoch of nucleosynthesis might raise the temperature of the universe and initiate further fusion processes (e.g. Dimopolous *et al.* 1988a,b). This requires the existence of a particle with the correct decay properties and it is not, in any case, clear whether this scenario can reproduce the consistency we have stressed above (e.g. Kolb & Turner 1990).

Probably the best argument for non-standard nucleosynthesis is the suggestion that the standard model itself may be flawed (e.g. Applegate & Hogan 1985; Malaney & Mathews 1993). For example, if the quark–hadron phase transition is a first-order transition then, as the universe cools, one would produce bubbles of the hadron phase inside the quark plasma (e.g. Bonometto & Pantano 1993). The transition proceeds only after the nucleation of these bubbles, and results in a very inhomogeneous distribution of hadrons with an almost uniform radiation background. In this situation, both protons and neutrons are strongly coupled to the radiation because of the efficiency of 'charged current' interactions. These reactions, however, freeze out at $T \simeq 1$ MeV so that the neutrons can then diffuse while the protons remain locked to the radiation field. The result of all this is that the n/p ratio, which is one of the fundamental determinants of the ^{4}He abundance, could vary substantially from place to place. In regions of relatively high proton density, every neutron will end up in a ^{4}He nucleus. In neutron-rich regions, however, the neutrons have to undergo β-decay before they can begin to fuse. The net result is less ^{4}He and more D than in the standard model, for the same value of η. The observed limits on cosmological abundances do not therefore in this case imply such a strong upper limit on Ω_b. It has even been suggested that such a mechanism may allow a critical density of baryons, $\Omega_b = 1$, to be compatible with observed element abundances. This idea is certainly interesting, but to find out whether it is correct one needs to perform a detailed numerical solution of the neutron transport and nucleosynthesis reactions, allowing for a strong spatial variation. In recent years, attempts have been made to perform such calculations but they have not been able to show convincingly that the standard model needs to be modified and the corresponding limits weakened.

In summary, then, there are a number of ways in which one might seek to 'refine' the standard model of nucleosynthesis, by introducing extra ingredients or relaxing the model assumptions. But the agreement of the standard model appears, at least for the time being, to be so adequate that one should perhaps avoid over-complicating things in this way. As always, however, one should bear in mind that these non-standard routes are by no means exhaustively ruled out.

4.4 Implications for Ω

4.4.1 *Summary of constraints*

The limits on Ω_b are of such importance – both historical and contemporary – that we repeat the main constraints here. Hedging our bets on whether the high deuterium values will stay or go away in due course, we offer two possible allowed ranges:

$$0.0052 < \Omega_b h^2 < 0.026, \tag{4.28}$$

if we take the very generous allowances for systematic errors in Copi *et al.* (1995a,b); or

$$0.0054 < \Omega_b h^2 < 0.007, \tag{4.29}$$

if we approach the high deuterium abundance as definitive, as in Rugers & Hogan (1996a,b).

Clearly, for any reasonable value of h the contribution of baryonic matter to the critical density is very small. If $h \simeq 0.4$ – unlikely, but perhaps just plausible – then (4.28) means that $\Omega_b < 0.16$; (4.29) means $\Omega_b < 0.044$. That nucleosynthesis favours a low value of the baryon density is not a new conclusion – indeed it was the subject of a classic paper of Gott *et al.* (1974). Today, however, many of the parameters needed for these calculations, such as the neutron half-life and the number of light neutrino types, are much better known from direct particle experiments.

4.4.2 Non-baryonic dark matter

The discrepancy between the values (4.28) and (4.29) and $\Omega_b = 1$ is the main argument for the existence of large amounts of non-baryonic dark matter (i.e. matter that is not involved in nuclear reactions) if the universe has a critical density. The problem is, however, that even if $0.1 < \Omega_0 < 0.3$, as indicated by various dynamical estimates we shall discuss later on in this book, one seems to require some of this to be non-baryonic† to be compatible with nucleosynthesis and the favoured values for h. Since the evidence for dark matter is in different forms in different contexts, we shall raise the issue again as we go along as to whether some or all of a given component can be in the form of baryons.

Of course, the only definitive way to answer the question of whether there does exist non-baryonic dark matter would be by direct detection of the particles from which it is made. We discussed some of the possibilities in §2.4.

4.5 Summary

The best present limits on the baryon density are those derived from comparison of nucleosynthesis data and theory, leading to (4.28) and (4.29). The main uncertainty arises from the as yet incomplete analysis of abundance data for distant objects. Taken with other estimates we present, these results are the prime reason to believe in the existence of non-baryonic matter in the universe.

As we have discussed, various possibilities for non-standard nucleosynthesis schemes do exist and it is possible that some of them can can accommodate a higher value of Ω_b, notably those in which inhomogeneities arise during the quark–hadron phase transition. Whether it is possible to accommodate $\Omega_b \simeq 1$ or even $\Omega_b = 0.2$ in such schemes is nevertheless unclear. In conclusion we would also like to suggest that, even if the standard model of nucleosynthesis is in accord with observations (which is quite remarkable, given the simplicity of the model), the constraints particularly on Ω_b emerging from these calculations are so fundamental to so many

† An alternative to non-baryonic material would be to have baryonic material form into primordial black holes before nucleosynthesis begins. To all intents and purposes, this kind of dark matter would act like CDM.

things that one should always keep an open mind about alternative, non-standard models which, as far as we are aware, are not completely excluded by observations. The theoretical analysis of such models needs to be pursued and refined, and we need also to extend the observational data (particularly at high redshifts) so as to clarify the situation. This is likely to remain the best way that will be available to us to determine the baryon abundance.

This issue, however, is an aside to the main question to which this book is addressed: that of the total value of Ω_0. As far as this issue goes, we can only make the very minimal statement that we at least have a fairly definite lower limit on $\Omega_0 h^2 \geq 0.005$ from nucleosynthesis considerations.

5
Astrophysical arguments

In this chapter we shall discuss the dark matter inferred from astrophysical† measurements. We divide these astrophysical arguments into three broad categories: galaxies, rich clusters of galaxies and the intergalactic medium. Because astrophysical processes (with the exception of gravitational effects) generally involve baryonic material only, the constraints we discuss frequently, though not exclusively, relate only to the baryonic contribution to the total density. We shall discuss constraints from large-scale structure in the matter distribution (i.e. clustering on scales greater than the scale of individual rich clusters) in the next chapter. For an extensive and influential survey of much of the astrophysical evidence see Peebles (1971), which serves as a standard reference for much that we discuss in this chapter; see also Faber & Gallagher (1979).

5.1 Galaxies

It was suggested as early as the 1930s that the total amount of matter in our own galaxy, the Milky Way, is greater than can be accounted for by the visible matter within it (e.g. Oort 1932). We shall not, however, go into any detail here concerning the evidence from stellar dynamics that there is dark matter in the disk of the Milky Way; see, for example, Bahcall (1984). This is still an open question. What we are interested in is the evidence for massive haloes of dark matter surrounding our own and other galaxies.

† By astrophysical, rather than cosmological, as in the previous chapter, we mean that the inferences discussed rely on local physical arguments pertaining to specific objects or phenomena rather than global constraints to do with cosmological models. As we shall see, however, the split is not all that clear: there is some overlap with the arguments from large-scale structure which we discuss in the next chapter.

5.1.1 The mass-to-light ratio

Before discussing the evidence for dark matter in galaxies and clusters, we need to introduce some notation. We begin by calculating the mean luminosity $\mathcal{L}_g$ per unit volume produced by galaxies, together with the mean value of M/L, the mass-to-light ratio, of the galaxies. Thus the mean density of material associated with galaxies can be written

$$\rho_{\rm g} = \mathcal{L}_{\rm g}\langle(M/L)\rangle. \tag{5.1}$$

The value of $\mathcal{L}_{\rm g}$ can be obtained from the *luminosity function* of the galaxies, $\Phi(L)$. This function is defined such that the number of galaxies per unit volume with luminosity in the range L to $L + {\rm d}L$ is given by

$${\rm d}N = \Phi(L){\rm d}L, \tag{5.2}$$

so that

$$\mathcal{L}_{\rm g} = \int_0^\infty \Phi(L)L{\rm d}L. \tag{5.3}$$

The best fit to the observed properties of galaxies is afforded by the *Schechter function*

$$\Phi(L) = \frac{\Phi_*}{L_*}\left(\frac{L}{L_*}\right)^{-\alpha} \exp\left(-\frac{L}{L_*}\right), \tag{5.4}$$

where rough values of the parameters are $\Phi_* \simeq 10^{-2}h^3$ Mpc^{-3}, $L_* \simeq 10^{10}h^{-2}\ L_\odot$ and $\alpha \simeq 1$ (e.g. Efstathiou, Ellis & Peterson 1988). The value of $\mathcal{L}_{\rm g}$ that results is

$$\mathcal{L}_{\rm g} \simeq 3.3 \times 10^8 h\ L_\odot{\rm Mpc}^{-3}. \tag{5.5}$$

This observed luminosity density can be used to determine the value that M/L must have in order for material associated with galaxies to produce a critical density universe. The result is that

$$(M/L)_{\rm crit} \simeq 1400h(M_\odot/L_\odot). \tag{5.6}$$

This figure serves as a useful yardstick against which to compare estimates of M/L from different systems. Before we discuss such estimates, however, we should make it clear that the following methods measure only the luminous mass directly associated with particular objects, and it is consequently quite possible for an observer to infer a smaller M/L within a given object than applies globally if there is smoothly distributed dark matter.

5.1.2 Spiral and elliptical galaxies

To derive the mass-to-light ratio M/L we must somehow measure the value of M. The best evidence for dark matter in galaxies and consequently high values of M/L comes from the behaviour of the rotation curves of spiral galaxies. Here one compares the observed curve with a theoretical model in which the rotation curve is produced by a distribution of gravitating material. There is strong evidence from 21 cm radio and optical observations that the rotation curves of spiral galaxies remain flat well outside the region in which most of the luminous material resides (e.g. Rubin, Ford & Thonnard 1980). A simple application of Newtonian gravity then suggests that spiral galaxies are embedded in quasi-spherical haloes with $\rho(r) \sim r^{-2}$ so that the mass contained within a given radius scales as $M \sim r$ and the rotation curve is consequently flat†. Typical values of the mass-to-light ratio inferred for such systems go up to $M/L \simeq 10h$ in solar units, with this rough figure dating back to Gott *et al.* (1974).

Persic & Salucci (1992) have examined the rotation curves of a sample of spiral galaxies in some detail and found evidence that the M/L ratio actually has a significant dependence on the intrinsic luminosity of the galaxy: $M/L \sim L^{0.35}$, suggesting that one should take a weighted average to get the global value of ρ_g corresponding to a value of

$$\Omega_{\rm spirals} \simeq 7 \times 10^{-4} \tag{5.7}$$

or an effective global value of $M/L \simeq 1$.

There is no direct evidence that elliptical and SO galaxies contain much dark matter in their inner regions, but X-ray observations of hot gas do suggest that these objects have dark haloes as well: see, for example, Sarazin (1986). Reviewing the evidence surrounding these objects, Persic & Salucci (1992) conclude that they contribute around

$$\Omega_{\rm ellipticals} \simeq 1.5 \times 10^{-3}. \tag{5.8}$$

† A more radical suggestion to account for the discrepancy between light profiles and rotation curves is to invoke large-scale departures from Newtonian Gravity; see, e.g., Milgrom (1983).

5.1.3 Galactic microlensing

Before carrying on with our inventory of the amount of dark matter inferred from galactic observations, we mention in passing the important constraints that gravitational lensing can place on the contribution of dark matter to the halo of our own Galaxy. Even if a gravitational lens is not strong enough to form two distinct images of a background source, as we discussed in Chapter 3, it may still amplify its brightness to an observable extent (e.g. Paczynski 1986a,b). The idea that galactic halo dark matter might lens the light from distant stars has recently borne fruit with convincing evidence for microlensing of stars in the Large Magellanic Cloud by sub-stellar mass objects in the halo of the Milky Way (Alcock *et al.* 1993; Auborg *et al.* 1993). Although these do not strongly constrain the total amount of dark matter in our Galaxy, the relatively small number of microlenses detected does constrain the contribution to the mass of the halo in brown dwarfs; see Carr (1994).

5.1.4 The galactic density of matter

Although the figures we have quoted above are open to some debate, they are the orders of magnitude generally derived from the mass-to-light ratios of matter directly related with the luminous content of the galaxies. However, the pitifully small contribution to Ω_0 that one obtains from these estimates can be increased if one assumes that the dark haloes extend far beyond the luminous extent of the galaxies. Reviewing the available evidence on the possible extension of galactic haloes, Bahcall, Lubin & Doorman (1995) have shown that if the halo extends to a radius R the global mass-to-light ratio is

$$M/L \simeq 60(R/100\,\mathrm{kpc})(M_\odot/L_\odot) \tag{5.9}$$

for spirals and

$$M/L \simeq 200(R/100\,\mathrm{kpc})(M_\odot/L_\odot) \tag{5.10}$$

for ellipticals. If one postulates that such haloes extend for the order of 1 Mpc then one is getting closer to the value needed to reconcile the matter associated with galaxies with that needed for a critical density universe. The behaviour of satellite galaxies suggests that haloes may indeed extend well beyond the visible limit of galaxies (Zaritsky *et al.* 1993). The direct evidence for having extremely extended haloes is, however, relatively slight: $r = 30h^{-1}$ kpc is the furthest one can go with total safety, suggesting the contribution to Ω from galactic haloes need be no more than

$$\Omega_{\mathrm{haloes}} \sim 0.02 \qquad (5.11)$$

(e.g. Trimble 1987). Notice that the directly inferred dark matter associated with galaxies is consistent with the nucleosynthesis bounds (4.28) but that, if the haloes are very extended, it appears some of the dark matter must be non-baryonic. Clearly it is important to extend our observations of the size of galactic haloes in the future.

5.2 Clusters of galaxies

Clusters of galaxies, particularly the very densest ones, have for many years provided strong evidence for the existence of more dark matter than is present in individual galaxies (e.g. Bahcall 1977). Indeed, the first evidence for extragalactic dark matter was found by Zwicky (1933) in an analysis of the nearest very large galaxy cluster, the Coma cluster. We shall illustrate the arguments in the following sections by using the Coma cluster as an example, though many other clusters have now been studied.

5.2.1 Galaxies in clusters

The most immediately obvious property of galaxy clusters is that they contain lots of galaxies. It is possible, therefore, to estimate the contribution to their total mass by counting galaxies in the manner indicated above and assigning a value of M/L. White *et al.* (1993) have shown that the total luminosity in blue light generated by stars in the galaxies within the central parts of the Coma cluster is $L \simeq 1.9 \times 10^{12} h^{-2} L_{\odot}$ and the the mass inferred is then

$$M_{\mathrm{gal}} \simeq 1.0 \times 10^{13} h^{-1} M_{\odot}. \qquad (5.12)$$

This is the mass within $r_A = 1.5h^{-1}$ Mpc of the centre of the cluster; r_A is called the *Abell radius.* As we shall see, this provides only a lower limit on the mass contained within the cluster.

5.2.2 Dynamical dark matter in clusters

The simplest kind of dynamical argument that can be used to estimate the total gravitating mass of a cluster of galaxies is based on the virial theorem. This method is particularly useful for rich clusters of galaxies like the Coma cluster, which one has reason to believe is a dynamically relaxed system.

The kinetic energy can be crudely estimated from the velocity dispersion of the galaxies in the cluster

$$E_k \simeq \frac{3}{2} M_{\rm tot} \sigma_1^2; \tag{5.13}$$

$M_{\rm tot}$ is the total mass of the cluster and $\sigma_1^2 = \langle v_r^2 \rangle^{1/2}$ is the line-of-sight velocity dispersion of the galaxies. The potential energy is given by

$$U \simeq -\frac{G M_{\rm tot}^2}{R_{\rm tot}} , \tag{5.14}$$

where $R_{\rm tot}$ is the gravitational radius of the cluster, which can be estimated from a model of its density profile, if one assumes that galaxies trace the same profile as the total mass.

A more sophisticated way of using the properties of equilibrium self-gravitating systems to infer their mass is to use the equation of hydrostatic equilibrium:

$$M(r) = -\frac{r\sigma^2(r)}{G} \left[\frac{\mathrm{d}\log\rho}{\mathrm{d}\log r} + \frac{\mathrm{d}\log\sigma_r^2}{\mathrm{d}\log r} + 2\beta \right] . \tag{5.15}$$

This gives the mass contained within a radius r in terms of the density profile $\rho(r)$ and the two independent velocity dispersions in the radial and tangential directions σ_r^2 and σ_t^2; the quantity

$$\beta = 1 - \frac{\sigma_t^2}{\sigma_r^2} \tag{5.16}$$

is a measure of the anisotropy of the radial velocity dispersion. In order to use this equation, one needs to know the profile of galaxies and velocity dispersion as a function of radius from the centre of the cluster. In reality, one can only measure the projected versions of these quantities, so the problem is formally indeterminate. One can, however, use a modelling procedure (e.g. The & White 1986) to perform an inversion of the projected profiles. For the Coma cluster, the result is a total dynamically inferred mass within the Abell radius of

$$M_{\rm tot} \simeq 6.8 \times 10^{14} h^{-1} M_\odot \qquad (5.17)$$

which corresponds to a value of $M/L \simeq 320h$. Galaxies therefore contribute only about 15 per cent of the mass of the Coma cluster. If this value of the mass-to-light ratio were global, then one would infer that

$$\Omega_{\rm clusters} \simeq 0.2 \qquad (5.18)$$

for the total density contributed by this form of dark matter. A recent study by Carlberg *et al.* (1996) of a sample of a further 16 galaxy clusters largely confirms these results: they find $\Omega_0 = 0.24 \pm 0.05$ for those components of the mass field in rich clusters. These authors suggest a possible systematic contribution to the errors of the order of 0.09 in Ω_0, so the actual value may be as low as $\Omega_0 \simeq 0.1$. Further studies should tighten this up.

If the value of M/L for galaxies were to be reconciled with the galactic value, one would have to have systematically overestimated the virial mass of the cluster. This might happen if the cluster were not gravitationally bound and virialised but instead were still freely expanding with the background cosmology. In such a case we would have

$$2E_k + U > 0 \qquad (5.19)$$

and, therefore, a smaller total mass. However, we would then expect the cluster to disperse on a characteristic time-scale $t_c \simeq l_c/\langle v^2\rangle^{1/2}$, where l_c is a representative length scale for the cluster and $\langle v^2\rangle^{1/2}$ is the root mean square peculiar velocity of the galaxies in the cluster; for the Coma cluster $t_c \simeq 1/16H_0$ and it is generally the case that t_c for clusters is much less than the Hubble time. If the clusters we observe were formed in a continuous fashion during the expansion of the universe, many such clusters must have already dispersed in this way. The space between clusters should therefore contain galaxies of the type usually found in clusters, i.e. elliptical and lenticular galaxies, and they might be expected to have large peculiar motions. One observes, however, that 'field' galaxies are usually spirals and they do not have particularly large peculiar velocities. It seems reasonable therefore to conclude that clusters must be bound objects.

It is also pertinent to question the assumption that the galaxy profile traces the dark matter: i.e. $\rho_g(r) \propto \rho_m(r)$. If the galaxies are less concentrated than the dark matter then one underestimates M/L, and vice versa. Theories of biased galaxy formation (Chapter 6) suggest that galaxies should be more concentrated than the total mass, which works in the direction of reconciling the value of M/L for Coma with the value needed for an $\Omega_0 = 1$ universe. It is important to develop these theories further to see how far this may go.

5.2.3 X-ray gas in clusters

Many rich clusters of galaxies are permeated by a tenuous gaseous atmosphere of X-ray emitting gas. Since the temperature and density profiles of the gas can be obtained with X-ray telescopes such as ROSAT and data on the X-ray spectrum of these objects is also often available, one can break the indeterminacy of the version of the hydrostatic equilibrium equation given above. The X-ray data also have the advantage that they are not susceptible to Poisson errors coming from the relatively small number of galaxies that exist at a given radius. Assuming the cluster is spherically symmetric and considering only the gaseous component, for simplicity, the equation of hydrostatic equilibrium becomes

$$M(r) = -\frac{k_B T(r) r}{G\mu m_p}\left[\frac{\mathrm{d}\log\rho}{\mathrm{d}\log r} + \frac{\mathrm{d}\log T}{\mathrm{d}\log r}\right]; \tag{5.20}$$

μ is the mean molecular weight of the gas. The procedure adopted is generally to use trial functions for $M(r)$ in order to obtain consistency with $T(r)$ and the spectrum data.

Good X-ray data from ROSAT have been used to model the gas distribution in the Coma cluster (Briel, Henry & Böhringer 1992) with the result that

$$M_{\mathrm{gas}} \simeq 5.5 \times 10^{13} h^{-5/2} M_\odot \tag{5.21}$$

for the mass inside the Abell radius. The gas contributes more than the galaxies, but is still less than the total mass.

A possible caveat exists here if there is a contribution to the hydrostatic balance from something other than the gas pressure. It has been suggested, for example, that an equipartition strength magnetic field could seriously interfere with this kind of estimate

(Loeb & Mao 1994), though there is no evidence for a field of the required strength in the Coma cluster. More realistically perhaps, one should note that the emissivity of the X-ray gas scales as n^2 where n is the number-density of ions, so an inhomogeneous distribution of gas could also bias the estimate.

5.2.4 The baryon catastrophe?

In an influential study, White *et al.* (1993) have analysed the contributions of X-ray gas, galaxies and dynamically inferred dark matter (discussed above) to the total mass of the Coma cluster. The fraction of the total mass contained in hot gas is embarrassingly large if one believes the ratio of baryonic mass to non-baryonic dark matter in Coma is similar to the global proportions required if Ω_b is consistent with the nucleosynthesis constraint, which they took to be given by equation (4.24). A strong constraint emerges:

$$\Omega_0 \leq \frac{0.15h^{-1/2}}{1 + 0.55h^{3/2}}. \tag{5.22}$$

Notice that this is well below unity for any likely value of h. This constraint is widely referred to as 'the baryon catastrophe', although it is only a catastrophe if you believe that $\Omega_0 \simeq 1$ in non-baryonic dark matter. The inferred limit on Ω_0 decreases further if one accepts the lower nucleosynthesis limit (4.29).

It is obviously an important future task to carry out similar analyses for clusters other than Coma to see if the baryon catastrophe is not just due to some peculiar property of the Coma cluster. There is, as yet, no evidence that this is the case.

5.2.5 Arcs, arclets and image distortions

There exists a possible independent test of the dynamical and X-ray masses of rich clusters, which does not depend on the assumption of virial or hydrostatic equilibrium. Gravitational lensing of the light from background objects depends on the total mass of the cluster whatever its form and physical state, leading to multiple and/or distorted images of the background object.

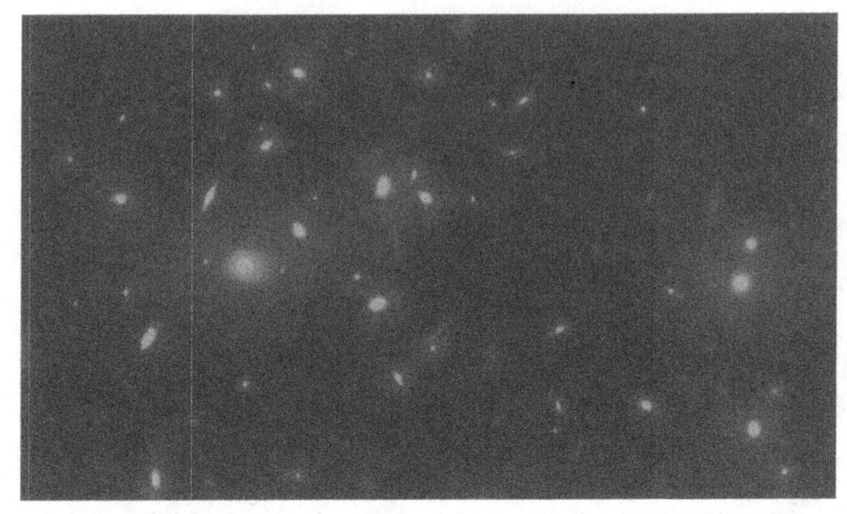

Fig. 5.1 HST image of the rich cluster Abell 2218 showing numerous giant arcs and arclets. Picture courtesy of Richard Ellis.

The possible lensing phenomena fall into two categories: *strong* lensing in rich clusters can probe the mass distribution in the central parts of these objects; *weak* lensing distortions of background galaxies can trace the mass distribution much further out from the cluster core. The discovery of giant arcs in images of rich clusters of galaxies as a manifestation of strong gravitational lensing (Tyson, Valdes & Wenk 1990; Fort & Mellier 1994; see Figure 5.1 for an example) has led to a considerable industry in using models of the cluster lens to determine the mass profile. The relevant parameter here is the Einstein radius b, which gives the rough angular size of arcs produced by a cluster with 3D velocity dispersion σ^2:

$$b \simeq \frac{4\pi\sigma^2}{c^2}\frac{D_{ls}}{D_os} \simeq 29'' \left(\frac{\sigma^2}{1000\text{km s}^{-1}}\right)\frac{D_{ls}}{D_{os}}, \tag{5.23}$$

where D_{ls} and D_{os} are the angular diameter distances from the lens to the background source and from the source to the observer respectively. Smaller arcs – usually called arclets – can be used to provide more detailed modelling of the lensing mass distribution. For recent applications of this idea, see Kneib *et al.* (1993) and Smail *et al.* (1995c); the latter authors, for example, infer a velocity dispersion of $\sigma^2 \simeq 1400$ km s^{-1} for the cluster AC114.

Important though these strong lensing studies undoubtedly are, they generally only probe the central parts of the cluster and say relatively little about the distribution of matter in the outskirts. They do, for example, seem to indicate that the total distribution of matter is more centrally concentrated than the gas distribution inferred from X-ray observations. On the other hand, estimates of the total masses obtained using strong lensing arguments are not in contradiction with virial analysis methods described above.

Weak lensing phenomena – the slight distortions of background galaxies produced by lines-of-sight further out from the cluster core – can yield constraints on the haloes of rich clusters (Kaiser & Squires 1993; Broadhurst, Taylor & Peacock 1995). It is also possible to use fluctuations in the $N(z)$ distribution of the galaxies behind the cluster to model the mass distribution (Smail *et al.* 1995a,b).

The technology of these methods has developed rapidly and has now been applied to several clusters. Preliminary results are generally indicative of a larger total mass than is inferred by virial arguments, suggesting that there exists even more dark matter than dynamics would suggest. For example, the rich cluster MS1224+20 seems to have a lensing mass of order three times the virial mass (Fahlman *et al.* 1994; Carlberg, Yee & Ellingson 1994); factors of 2 or more appear to be commonplace in other clusters. These determinations are working in the direction of consistency between cluster masses and $\Omega_0 \simeq 1$, but not so far by a large enough factor. In any case this technique is relatively young and it is possible that not all the systematic errors have yet been ironed out, so we take these results as indicating that this is a good – indeed important – method for use in future studies, rather than one which is providing definitive results at the present time.

5.2.6 Subclustering and time-scales

Another aspect of rich clusters that can be used to probe the cosmological density is their intrinsic 'clumpiness'. The fact that at least some clusters appear to contain substructure (e.g. West, Jones & Forman 1995) and are not therefore completely relaxed systems can be used to estimate how many such clusters have formed recently. Newly formed clusters are dynamically young and can therefore be expected to exhibit substructure; early-forming clusters are dynamically relaxed and should have smooth density profiles. Cluster formation is expected to occur systematically earlier in a low density universe compared to one with $\Omega_0 \simeq 1$, so the fraction of clusters exhibiting substructure is expected to depend on Ω_0.

Attempts have recently been made to translate this qualitative argument into a quantitative constraint on Ω_0. The analysis depends on approximating the collapse of a cluster by the equations describing the collisionless evolution of self-gravitating systems. The original analysis (Richstone, Loeb & Turner 1992) assumed spherical collapse, but this has been generalised to apply also to anisotropic initial proto-cluster configurations (Bartellman, Ehlers & Schneider 1993) and the more general context of hierarchical structure formation theories (Lacey & Cole 1993). If one accepts that a significant fraction of clusters ($\sim 30\%$) show substructure then this analysis seems to require $\Omega_0 \geq 0.5$; see also Mohr *et al.* (1995). On the other hand, one has to be sure that one adopts an objective measure of substructure that can be applied to the real cluster data and to theoretical calculations to avoid comparing apples with oranges (e.g. Tsai & Buote 1996). Wilson, Cole & Frenk (1996) have suggested that many of these uncertainties can be avoided by using the cluster mass profile inferred from gravitational lensing considerations to check for substructure in the distribution of gravitating mass: simulations seem to indicate that this can be done reliably and that strong constraints will emerge from future studies of clusters of galaxies.

These are indeed interesting arguments. However, one must bear in mind that whilst a collisionless collapse may be a good model for what happens if the cluster is formed mainly from weakly interacting dark matter, there is much less motivation for such a model if a sizeable fraction of the cluster mass is in the hot gas component as seems to be the case. Hydrodynamical effects may be dominant in the latter situation and even magnetic fields may play a role in maintaining the substructure. To an extent, therefore, it could be argued that this analysis assumes what it set out to prove, and therefore does not strongly constrain the alternative, low-density model. Notice also that if there is evidence for substructure in clusters then that is reason to caution against using virial arguments to determine their masses: it is unlikely that virial equilibrium holds in the presence of significant substructure (White, Briel & Henry 1993), thus perhaps explaining the (possible) discrepancy between virial and lensing arguments. Indeed it is not entirely clear that substructure cannot be generated during cluster collapse by the action of a gravi-thermal instability (Lynden-Bell & Wood 1968; Padmanabhan 1990).

5.3 The intergalactic medium (IGM)

We now turn our attention to various constraints imposed on the contribution of the matter intervening between galaxies and clusters by various categories of observations. These *all* constrain only the baryonic contribution to Ω_0 and, for the most part, only part of that. For more details, see Peebles (1971), Peebles (1993) and Coles & Lucchin (1995).

5.3.1 Quasar spectra

Observations of quasar spectra allow one to probe a line of sight from our galaxy to the quasar. Absorption or scattering of light during its journey to us can, in principle, be detected by its effect upon the spectrum of the quasar. This, in turn, can be used to constrain the number and properties of absorbers or scatterers, which, whatever they are, must be associated with the baryonic content of the IGM. Before we describe the possibilities, it is therefore useful to write down the mean number density of baryons as a function of Ω_b and redshift:

$$n_b \simeq 1.1 \times 10^{-5} \Omega_b h^2 (1+z)^3 \text{ cm}^{-3}. \qquad (5.24)$$

This is an important reference quantity for the following considerations.

The Gunn–Peterson test

Neutral hydrogen has a resonant scattering feature associated with the Lyman-α atomic transition. This resonance is so strong that it is possible for a relatively low neutral hydrogen column density (i.e. number-density per unit area of atoms, integrated along the line of sight) to cause a significant apparent absorption at the appropriate wavelength for the transition. Let us suppose that light travels towards us through a uniform background of neutral hydrogen. The optical depth for scattering is

$$\tau(\lambda_0) = \frac{c}{H_0} \int \sigma(\lambda_0 a/a_0) n_I(t) \Omega^{-1/2} \left(\frac{a_0}{a}\right)^{-3/2} \frac{da}{a} , \qquad (5.25)$$

where $\sigma(\lambda)$ is the cross-section at resonance and n_I is the proper density of neutral hydrogen atoms at the redshift corresponding to this resonance. (The usual convention is that HI refers to neutral and HII to ionised hydrogen.) We have assumed in this equation that the universe is matter dominated. The integral is taken over the width of the resonance line (which is very narrow and can therefore be approximated by a delta function) and yields a result for τ at some observed wavelength λ_0. It therefore follows that

$$\tau = \frac{3\Lambda\lambda_\alpha^3 n_I}{8\pi H_0 \Omega^{1/2}} (1+z)^{-3/2}, \qquad (5.26)$$

where $\Lambda = 6.25 \times 10^8$ s^{-1} is the rate of spontaneous decays from the 2p to 1s level of hydrogen (the Lyman-α emission transition); λ_α is the wavelength corresponding to this transition, i.e. 1216 Å. The equation (5.26) can be inverted to yield

$$n_I = 2.4 \times 10^{-11}\ \Omega^{1/2} h (1+z)^{3/2} \tau\ \mathrm{cm}^{-3} . \qquad (5.27)$$

This corresponds to the optical depth τ at $z = (\lambda_0/\lambda_\alpha) - 1$, when observed at a wavelength λ_0.

The *Gunn–Peterson test* takes note of the fact that there is no apparent drop between the long wavelength side of the Lyman-α emission line in quasar spectra and the short wavelength side, where extinction by scattering might be expected (Gunn & Peterson 1965). Observations suggest a (conservative) upper limit on τ of the order of 0.1, which translates into a very tight bound on n_I:

$$n_I < 2 \times 10^{-12}\ \Omega^{1/2} h (1+z)^{3/2}\ \mathrm{cm}^{-3}. \qquad (5.28)$$

Comparing this with equation (5.24) with $\Omega_b = 1$ yields a constraint on the contribution to the critical density owing to neutral hydrogen:

$$\Omega(n_I) < 2 \times 10^{-7}\ \Omega^{1/2} h^{-1} (1+z)^{-3/2}. \qquad (5.29)$$

There is no alternative but to assume that, by the epoch one can probe directly with quasar spectra (which corresponds to $z \simeq 4$), the density of any uniform neutral component of the IGM was very small indeed.

One can translate this result for the neutral hydrogen into a constraint on the plasma density at high temperatures by considering the balance between collisional ionisation reactions,

$$\mathrm{H} + e^- \rightarrow p + e^- + e^- , \tag{5.30}$$

and recombination reactions of the form

$$p + e^- \rightarrow H + \gamma. \tag{5.31}$$

The physics of this balance is complicated by the fact that the cross-sections for these reactions are functions of temperature. It turns out that the ratio of neutral hydrogen to ionised hydrogen, n_I/n_{II}, has a minimum at a temperature around 10^6 K, and at this temperature the equilibrium ratio is

$$\frac{n_I}{n_{II}} \simeq 5 \times 10^{-7}. \tag{5.32}$$

Since this is the minimum possible value, the upper limit on n_I therefore gives an upper limit on the total density in the IGM which we can assume to be made entirely of hydrogen:

$$\Omega_{\mathrm{IGM}} < 0.4\Omega^{1/2}h^{-1}(1+z)^{-3/2}. \tag{5.33}$$

If the temperature is much lower than 10^6 K, the dominant mechanism for ionisation could be electromagnetic radiation. In this case one must consider the equilibrium between radiative ionisation and recombination, which is more complex and requires some assumptions about the ionising flux. There are probably enough high energy photons from quasars at around $z \simeq 3$ to ionise most of the baryons if the value of Ω_b is not near unity, and there is also the possibility that early star formation in protogalaxies could also contribute substantially. Another complication is that the spatial distribution of the IGM might be clumpy, which alters the average rate of recombination reactions but not the mean rate of ionisations. One can show that, for temperatures around 10^4K, the constraint emerges that

$$\Omega_{\mathrm{IGM}} < 0.4 I_{21}\Omega^{1/2}h^{-3/2}(1+z)^{9/4}, \tag{5.34}$$

if the medium is not clumpy and the ionising flux, I_{21}, is measured in units of 10^{-21} erg cm^{-2} s^{-1} Hz^{-1} ster^{-1}. The limit is reduced if there is a significant clumping of the gas.

These results suggest that the total IGM density cannot have been more than $\Omega_{\mathrm{IGM}} \simeq 0.03$ at $z \simeq 3$, whatever the temperature of the plasma. This limit is compatible with the nucleosynthesis bounds given in §8.6.

Absorption line systems

Although quasar spectra do not exhibit any general absorption consistent with a smoothly distributed hydrogen component, there are many absorption lines in such spectra which are interpreted as being due to clouds intervening between the quasar and the observer and absorbing at the Lyman-α resonance. Note that the recent claimed detections of deuterium in quasar spectra exploit the same phenomenon, except at a slightly different frequency because of the difference in nuclear mass between hydrogen and deuterium.

The clouds are grouped into three categories depending on their column density, which can be obtained from the strength of the absorption line. The strongest absorbers have column densities $\Sigma \simeq 10^{20}$ atoms cm^{-2} or more, which are comparable to the column densities of interstellar gas in a present-day spiral galaxy. This is enough to produce a very wide absorption trough at the Lyman-α wavelength, and these systems are usually called *damped Lyman-α systems.* These are relatively rare, and are usually interpreted as being the progenitors of spiral disks. They occur at redshifts up to around 3. A more abundant type of object is the *Lyman limit system.* These have $\Sigma \simeq 10^{17}$ atoms cm^{-2} and are dense enough to block radiation at wavelengths near the photoionisation edge of the Lyman series of lines. Smaller features, with $\Sigma \simeq 10^{14}$ atoms cm^{-2}, appear as sharp absorption lines at the Lyman-α wavelength. These are very common, and reveal themselves as a 'forest' of lines in the spectra of quasars, hence the term *Lyman-α forest.* The importance of the Lyman limit is that, at this column density, the material at the centre of the cloud will be shielded from ionising radiation by the material at its edge. At lower densities this cannot happen.

The damped Lyman-α systems have surface densities similar to spiral disks. It is natural therefore to interpret them as protogalactic disks. The only problem with this interpretation is that there are about ten times as many such systems at $z \simeq 3$ than one would expect by extrapolating backwards the present number of spiral galaxies. This may mean that, at high redshift, these galaxies are surrounded by gas clouds or very large neutral hydrogen disks which get destroyed as the galaxies evolve. It may also be that many of these objects end up as low surface brightness galaxies at the present epoch which do not form stars very efficiently: in such a case the present number of bright spirals is an underestimate of the number of damped Lyman-α systems that survive to the present epoch. It is also pertinent to mention that these systems have also been detected in CaII, MgII or CIV lines and that they do seem to have significant abundances of elements heavier than helium. There is some evidence that the fraction of heavy elements decreases at high redshifts (Wolfe 1993). We discuss the constraints on structure formation theories that can be imposed by such systems in the next chapter.

The Lyman-α forest clouds have a number of interesting properties. For a start they provide evidence that quasars are capable of ionising the IGM. The number densities of systems towards different quasars are similar, which strengthens the impression that they are intervening objects and not connected with the quasar. At redshifts near that of the quasar the number density decreases markedly, an effect known as the *proximity effect*. The idea here is that radiation from the quasar substantially reduces the neutral hydrogen fraction in the clouds by ionisation, thus inhibiting absorption at the Lyman-α resonance. Secondly, the total mass in the clouds appears to be close to that in the damped systems or that seen in present-day galaxies. This would be surprising if the forest clouds were part of an evolving clustering hierarchy, but if they almost fill space then one might not see any strong correlations in any case. Thirdly, the comoving number density of such systems is changing strongly with redshift, indicating, perhaps, that the clouds are undergoing dissipation. Finally, and most interestingly from the point of view of structure formation, the absorption systems seem to be unclustered, in contrast to the distribution of galaxies. How these smaller Lyman-α systems fit into a picture of galaxy formation is at present unclear.

Summary

In summary we can see that very strong upper limits can be placed on the amount of neutral hydrogen as a fraction of the critical density from the Gunn–Peterson test. Allowing for some ionising radiation at high redshift makes it possible to show compatibility of the inferred total baryon density with the nucleosynthesis bounds discussed in the previous chapter. Lyman-α absorption systems give even less direct information, except that the amount of gas contained in such systems might lead to some constraints on the fraction of baryons that become confined in bound structures at high redshift and hence give indirect evidence concerning Ω_0.

5.3.2 Spectral distortions of the CMB

The extremely accurate black-body shape of the spectrum of the cosmic microwave background (Mather *et al.* 1994) can furnish constraints on the ionised component of the IGM. Compton scattering of CMB photons by ionised material is expected to distort the shape of the spectrum in a way that depends upon the properties of the ionised medium. This effect is known as the *Sunyaev–Zel'dovich effect* (Sunyaev & Zel'dovich 1969). The relevant parameter is the so-called y-parameter:

$$y = \int_{t_{\rm min}}^{t_{\rm max}} \frac{k(T_e - T_r)}{m_e c^2} \sigma_t n_e(z) c dt, \qquad (5.35)$$

where the integral is taken over the time the photon takes to traverse the ionised medium. In most circumstances only one parameter is relevant, because the electron temperature T_e is much greater than the radiation temperature T_r.

When CMB photons scatter through material which has been heated in this way the shape of the spectrum is distorted in both the Rayleigh–Jeans and Wien regions. If $y < 0.25$ the shape of the Rayleigh–Jeans part of the spectrum does not change, but the effective temperature changes according to $T = T_r \exp(-2y)$. At high frequencies the intensity actually increases. This is due to low-frequency CMB photons being boosted in energy by Compton scattering and transferred to high-frequency parts of the spectrum. The Sunyaev–Zel'dovich effect allows one to place constraints on the properties of the intergalactic medium. If the hot gas is smoothly distributed, then one would not expect to see any angular variation in the temperature of the CMB radiation as a result of this phenomenon. However, the Sunyaev–Zel'dovich effect is frequency dependent: the dip associated with clusters appears in the

Rayleigh–Jeans region of the CMB spectrum. If one measures this spectrum one would expect a smooth gas distribution to produce a distortion of the black-body shape owing to scattering as the CMB photons traverse the IGM. The same will happen if gas is distributed in objects at high redshift which are too distant to be resolved. The importance of this effect has been emphasised by the CMB spectrum observed by the FIRAS experiment on COBE, which has placed the constraint $y < 3 \times 10^{-5}$ (Mather *et al.* 1994).

From equation (5.35) the contribution to y from a hot plasma with mean pressure $n_e k_B T_e$ at a redshift z is

$$y \simeq \sigma_T n_e ct \frac{k_B T_e}{m_e c^2} , \tag{5.36}$$

where the suffix e refers to the electrons. Various kinds of object containing hot gas could, in principle, contribute significantly to y. If Lyman-α clouds are in pressure balance at $z \simeq 3$, then they will contribute only a small fraction of the observational limit on y, so these clouds are unlikely to have an effect on the CMB spectrum. Similarly, if galaxies form at high redshifts with circular velocities v, then one can write

$$y \simeq \sigma_T n_e ct \left(\frac{v}{c}\right)^2 , \tag{5.37}$$

which is, approximately,

$$y \simeq 10^{-8} h \Omega_g \Omega^{-1/2} (1+z)^{3/2} \tag{5.38}$$

if $v \simeq 100$ km s^{-1} and Ω_g is the fractional contribution of hot gas to the critical density. The contribution from rich clusters is similarly small because, as we have seen, the gas in these objects only contributes around $\Omega_g \simeq 0.003$. On the other hand, a smooth hot IGM can have a significant effect on y, as we shall see in the next section.

5.3.3 A hot IGM?

A hot plasma produces radiation through thermal bremsstrahlung. The luminosity density at a frequency ν produced by this process for a pure hydrogen plasma is given approximately by

$$J(\nu) = 5.4 \times 10^{-39} n_e^2 T_e^{-1/2} \exp(-h\nu / k_B T_e), \tag{5.39}$$

in c.g.s. units, i.e. erg cm^{-3} s^{-1} ster^{-1} Hz^{-1}. The integrated background observed now at a frequency ν is

$$I(\nu) = \int cJ(\nu a_0/a, t)(a/a_0)^3 dt, \tag{5.40}$$

where the integral is taken over a line of sight through the medium. If the emission takes place predominantly at a redshift z, then

$$I(\nu) = 4 \times 10^{-23} \left(\frac{T_e}{10^4 \text{ K}}\right)^{-1/2} \frac{h^3 \Omega_{\text{IGM}}}{\Omega^{1/2}} (1+z)^{3/2} \tag{5.41}$$

for $h\nu \ll k_B T_e$, again in c.g.s. units: erg cm^{-2} s^{-1} ster^{-1} Hz^{-1}.

It has been known for some time that there exists a smooth background of X-ray emission (e.g. Boldt 1987). It is not known at present precisely what is responsible for this background but many classes of object can, in principle, contribute. Clusters of galaxies, quasars and active galaxies at high redshift and even starburst galaxies at relatively low redshift might be significant contributors to it. Disentangling these components is difficult and we shall not attempt to do it here. When the nature of the background is clarified, its spectrum and anisotropy may well provide strong constraints on models for the origin of quasars and other high-redshift objects. We shall concentrate on the constraints this background imposes on the IGM. The present surface brightness of the X-ray background is

$$I(\nu) \simeq 3 \times 10^{-26} \text{ erg cm}^{-3} \text{ s}^{-1} \text{ ster}^{-1} \text{ Hz}^{-1} \tag{5.42}$$

at energies around 3 keV. Suppose a fraction f of this is produced by a hot IGM with temperature $T \simeq 10^8(1+z)$ K; in this case

$$\Omega_{\text{IGM}} \simeq 0.3 f\, \Omega^{1/4} h^{-3/2} \left(\frac{T}{10^8 \text{ K}}\right)^{1/4} (1+z)^{-1/2}, \tag{5.43}$$

so that, if the plasma is smooth, the y-parameter is

$$y \simeq 2 \times 10^{-4} (1+z)^2 \left(\frac{f}{h\Omega^{1/2}}\right)^{1/2}. \tag{5.44}$$

These considerations here basically restrict the amount of very hot gas that can be distributed in a smooth component on cosmological scales. If the plasma were hot enough to produce a large fraction of the X-ray background it would also violate the constraints on the y distortion of the CMB spectrum. There is still plenty of freedom to have an intergalactic component which is not hot enough to produce the X-ray background but still satisfy the

Gunn–Peterson test discussed above, so the implications of this result for the value of Ω_0 are indeed slight, but they do at least provide relatively direct constraints on the behaviour of at least part of the baryonic component, and these are not inconsistent with the other arguments we have discussed.

5.4 Summary

We end this chapter by summarising the astrophysical arguments we have presented and their implication for Ω_0.

• The directly inferred dark matter in galaxies contributes up to around half a per cent of the critical density. This can be increased to around two per cent if one assumes the haloes extend beyond about $30h^{-1}$ kpc, and more if the haloes go further than that.
• Cluster masses obtained using virial arguments and X-ray observations yield mass-to-light ratios as large as $M/L \sim 200$–400. If these are representative of the total mass-to-light ratio pertaining to the whole universe then one infers $\Omega_0 \simeq 0.1$–0.3.
• If the standard nucleosynthesis bounds apply, not all of the dark matter in galaxies, or in extended haloes, can be baryonic.
• The baryon fraction in clusters appears to be higher than can be accommodated in models with $\Omega_0 = 1$ and in which the nucleosynthesis limits on Ω_b are satisfied. This is evidence in favour of a low total $\Omega_0 \simeq 0.2$.
• Strong gravitational lensing phenomena constrain the mass profile in the inner regions of clusters, while weak lensing arguments give more direct constraints on the total cluster mass. Present results seem to indicate higher total masses than is inferred from dynamical estimates but there are still quite large errors in the observations.
• The existence of substructure in clusters provides some circumstantial evidence for high values of $\Omega_0 \simeq 0.5$, but this depends on the particular way in which cluster formation is modelled and is therefore rather more indirect than the other constraints we have discussed.
• Astrophysical arguments can constrain the amount of neutral and ionised gas contained within the intergalactic medium. The principal results are that (i) the IGM was ionised at $z \simeq 4$ and

that (ii) not all of the X-ray background can be produced by a smooth hot IGM. These arguments do not, however, constrain the global value of Ω_0.

All these arguments need further development, but the most exciting and promising is that due to gravitational lensing. The observations here are improving rapidly and will give important data in the near future. The theoretical challenge is to reduce the model dependence of the analysis of this data. Clearly it will also be important to extend the analyses of X-ray data from clusters, to check whether the limits determined so far for the Coma cluster are, in fact, as strong evidence for a low value of Ω_0 as it appears at the moment.

6
Large-scale structure

We now turn our attention to the evidence from observations of galaxy clustering and peculiar motions on very large scales. In recent years this field has generated a large number of estimates of Ω_0, many of which are consistent with unity. Since these studies probe larger scales than the dynamical measurements discussed in Chapter 5, one might be tempted to take the large-scale structure as providing truer indications of the cosmological density of matter. On the other hand, it is at large scales that accurate data are hardest to obtain. Moreover, very large scale structures are not fully evolved dynamically, so one cannot safely employ equilibrium arguments in this case. The result is that one is generally forced to employ simplified dynamical arguments (based on perturbation theory), introduce various modelling assumptions into the analysis, and in many cases adopt a statistical approach. The global value of Ω_0 is just one of several parameters upon which the development of galaxy clustering depends, so results are likely to be less direct than obtained by other approaches. Moreover, it may turn out that the gravitational instability paradigm, which forms the basis of the discussion in this chapter, is not the right way to talk about structure formation. Perhaps some additional factor, such as a primordial magnetic field (Coles 1992) plays the dominant rolc. Nevertheless, there is a persuasive simplicity about the standard picture and it seems to accommodate many diverse aspects of clustering evolution, so we shall accept it for the sake of this argument.

6.1 Theoretical prelude

We begin our discussion of the evidence from large-scale structure with a brief introduction to theoretical models of the formation and evolution of cosmic structure. For more details, see, e.g., Coles & Lucchin (1995).

6.1.1 Gravitational instability

The equations governing the expansion of a homogeneous and isotropic universe were introduced in Chapter 1. We now need to consider the more difficult problem of dealing with the evolution of inhomogeneities. We are helped in this task by the fact that we expect such inhomogeneities to be of very small amplitude early on so we can adopt a kind of perturbative approach, at least for the early stages of the problem. The way to proceed is to perturb the Robertson–Walker metric given by equation (1.1), to represent small fluctuations in the geometry and hence the matter distribution through the Einstein equations. An appropriate choice for the line element which describes the perturbed FRW universe is

$$\mathrm{d}s^2 = \mathrm{d}s^2_{FRW} + a^2(t)h_{\alpha\beta}\mathrm{d}x^\alpha \mathrm{d}x^\beta, \tag{6.1}$$

where $\alpha, \beta = 1, 2, 3$, and ds^2_{FRW} is the FRW line element given by equation (1.1).

In this 'linearised' approximation, perturbations of the metric are assumed to be small ($h_{\alpha\beta}h^{\alpha\beta} \ll 1$), and the length scale of the perturbations is always much smaller than the effective cosmological horizon c/H_0, so that a Newtonian treatment of the subject is expected to be valid. If the mean free path of a particle is small, matter can be treated as an ideal fluid and the Newtonian equations governing the motion of gravitating particles in an expanding universe can be written in terms of $\mathbf{x} = \mathbf{r}/a$ (the comoving spatial coordinate, which is fixed for observers moving with the Hubble expansion), $\mathbf{v} = \dot{\mathbf{r}} - H\mathbf{r} = a\dot{\mathbf{x}}$ (the peculiar velocity field, representing departures of the matter motion from pure Hubble expansion), $\phi(\mathbf{x}, t)$ (the peculiar Newtonian gravitational potential, i.e. the fluctuations in potential with respect to the homogeneous background) and $\rho(\mathbf{x}, t)$ (the matter density). Using these variables we obtain, first, *the Euler equation*:

$$\frac{\partial(a\mathbf{v})}{\partial t} + (\mathbf{v} \cdot \nabla_{\mathbf{x}})\mathbf{v} = -\frac{1}{\rho}\nabla_{\mathbf{x}}p - \nabla_{\mathbf{x}}\phi \, . \tag{6.2}$$

The second term on the right-hand side of equation (6.2) is the peculiar gravitational force, which can be written in terms of $\mathbf{g} = -\nabla_{\mathbf{x}}\phi/a$, the peculiar gravitational acceleration of the fluid element. If the velocity flow is irrotational, $\mathbf{v}$ can be rewritten in terms of a velocity potential ϕ_v: $\mathbf{v} = -\nabla_{\mathbf{x}}\phi_v/a$. Next we have the *continuity equation*:

$$\frac{\partial \rho}{\partial t} + 3H\rho + \frac{1}{a}\nabla_{\mathbf{x}}(\rho\mathbf{v}) = 0, \tag{6.3}$$

which expresses the conservation of matter, and finally the *Poisson equation*:

$$\nabla_{\mathbf{x}}{}^2\phi = 4\pi G a^2(\rho - \rho_0) = 4\pi G a^2 \rho_0 \delta, \tag{6.4}$$

describing Newtonian gravity. Here ρ_0 is the mean background density, and

$$\delta \equiv \frac{\rho - \rho_0}{\rho_0} \tag{6.5}$$

is the *density contrast.*

6.1.2 Linear perturbation theory

The next step is to linearise the Euler, continuity and Poisson equations by perturbing physical quantities defined as functions of Eulerian coordinates, i.e. relative to an unperturbed coordinate system. Expanding ρ, $\mathbf{v}$ and ϕ perturbatively and keeping only the first-order terms in equations (6.2) and (6.3) gives the linearised continuity equation:

$$\frac{\partial \delta}{\partial t} = -\frac{1}{a}\nabla_{\mathbf{x}} \cdot \mathbf{v}, \tag{6.6}$$

which can be inverted, with a suitable choice of boundary conditions, to yield

$$\delta = -\frac{1}{aHf}\left(\nabla_{\mathbf{x}} \cdot \mathbf{v}\right). \tag{6.7}$$

The factor aH is simply a scaling of units and can be avoided by measuring distances in units of km s^{-1}: the 'observable' here is the function f, given by

$$f \equiv \frac{d\log\delta}{d\log a} \simeq \Omega_0^{0.6} + \frac{\Omega_\Lambda}{70}\left(1 + \frac{\Omega_0}{2}\right), \tag{6.8}$$

where $\Omega_\Lambda = \Lambda/3H_0^2$ is the effective fraction of the critical energy density contributed by a cosmological constant term; the right-hand side of this equation is simply a fitting formula to a numerical solution of equation (6.11) below. This form of the continuity equation allows one, at least in principle, to determine f by comparing peculiar velocities with density or gravitational potential fluctuations inferred from galaxy counts, a possibility we shall discuss again in §6.4.

The linearised Euler and Poisson equations are

$$\frac{\partial \mathbf{v}}{\partial t} + \frac{\dot{a}}{a}\mathbf{v} = -\frac{1}{\rho a}\nabla_{\mathbf{x}} p - \frac{1}{a}\nabla_{\mathbf{x}}\phi, \tag{6.9}$$

$$\nabla_{\mathbf{x}}{}^2\phi = 4\pi G a^2 \rho_0 \delta; \tag{6.10}$$

$|\mathbf{v}|, |\phi|, |\delta| \ll 1$ in equations (6.7), (6.9) & (6.10). From these equations, and if one ignores pressure forces, it is easy to obtain an equation for the evolution of δ:

$$\ddot{\delta} + 2H\dot{\delta} - \frac{3}{2}\Omega H^2 \delta = 0. \tag{6.11}$$

For a spatially flat universe dominated by pressureless matter, $\rho_0(t) = 1/6\pi G t^2$ and equation (6.11) admits two linearly independent power law solutions $\delta(\mathbf{x}, t) = D_\pm(t)\delta(\mathbf{x})$, where $D_+(t) \propto a(t) \propto t^{2/3}$ is the growing mode and $D_-(t) \propto t^{-1}$ is the decaying mode. The growing mode in an expanding universe increases as a power of t whereas Jeans' instability in a static background has an exponentially fast fluctuation growth. The effect of the expansion of the universe is to slow down the rate at which material can accrete onto the primordial 'seed' fluctuations.

From equations (6.7)–(6.10), still ignoring pressure, we also get

$$\mathbf{v} = -\frac{2f}{3\Omega H a}\nabla_{\mathbf{x}}\phi + \frac{\mathrm{const}}{a(t)}, \tag{6.12}$$

which demonstrates that the velocity flow associated with the growing mode in the linear regime is curl-free, as it can be expressed as the gradient of a scalar potential function†: this is the basis of a method for analysing peculiar velocities which we shall discuss below.

† See equation (6.34) below for an alternative expression.

6.1.3 Primordial density fluctuations

The above considerations apply to the evolution of a single Fourier mode of the density field $\delta(\mathbf{x},t) = D_+(t)\delta(\mathbf{x})$. What is more likely to be relevant, however, is the case of a superposition of waves, resulting from some kind of stochastic process. In the case of a statistically homogeneous distribution, i.e. a distribution whose statistical properties are the same everywhere, the Fourier modes are independent, as they are eigenfunctions of the spatial translation operator $i\nabla_{\mathbf{x}}$. The density field will then consist of a stochastic superposition of such modes with different amplitudes. A statistical description of the initial perturbations is therefore required, and any comparison between theory and observations will also have to be statistical.

The spatial Fourier transform of $\delta(\mathbf{x})$ is

$$\hat{\delta}(\mathbf{k}) = \frac{1}{(2\pi)^3}\int \mathrm{d}^3\mathbf{x} e^{-i\mathbf{k}\cdot\mathbf{x}}\delta(\mathbf{x}), \tag{6.13}$$

which has inverse

$$\delta(\mathbf{x}) = \int \mathrm{d}^3\mathbf{k} e^{i\mathbf{k}\cdot\mathbf{x}}\hat{\delta}(\mathbf{k}). \tag{6.14}$$

In the light of the preceding comments, it is useful to specify the properties of δ in terms of $\hat{\delta}$. We can define the *power-spectrum* of the field to be (essentially) the variance of the amplitudes at a given value of $\mathbf{k}$:

$$\langle\hat{\delta}(\mathbf{k}_1)\hat{\delta}(\mathbf{k}_2)\rangle = P(k_1)\delta^D(\mathbf{k}_1 + \mathbf{k}_2), \tag{6.15}$$

where δ^D is the Dirac delta function; this rather cumbersome definition takes account of the translation symmetry and reality requirements for $P(k)$; isotropy is expressed by $P(\mathbf{k}) = P(k)$. The analogous quantity in real space is called the two-point correlation function or, more correctly, the autocovariance function, of $\delta(\mathbf{x})$:

$$\langle\delta(\mathbf{x}_1)\delta(\mathbf{x}_2)\rangle = \xi(|\mathbf{x}_1 - \mathbf{x}_2|) = \xi(\mathbf{r}) = \xi(r), \tag{6.16}$$

which is related to the power spectrum via a Fourier transform (a fact known as the Wiener–Khintchin theorem).

The power-spectrum or, equivalently, the autocovariance function, of the density field is particularly important because it provides a complete statistical characterisation of a particular kind of stochastic process: a *Gaussian random field.* This class of field is the generic prediction of inflationary models, in which the density perturbations are generated by quantum fluctuations (e.g. Brandenberger 1985) in a scalar field during the inflationary epoch. One also might expect Gaussian fluctuations on more general grounds, as such fluctuations are generically produced by linear physical processes as a consequence of the central limit theorem. We shall assume in this paper that the primordial $\delta(\mathbf{x})$ is a Gaussian random field. Formally, this means that all the joint probability distributions of $\delta(\mathbf{x}_1)\ldots\delta(\mathbf{x}_n)$ are n-variate Gaussian distributions for this type of field, and this property allows many properties of the spatial distribution of density fluctuations to be calculated analytically (e.g. Bardeen *et al.* 1986).

Many other physically interesting properties can be related to the power-spectrum. For example, consider the mean square density fluctuation at a point

$$\langle \delta^2 \rangle = 4\pi \int_0^\infty P(k) k^2 \mathrm{d}k. \tag{6.17}$$

This definition of the mean square fluctuation takes into account contributions from all length scales and may, depending on the form of $P(k)$, actually diverge. A more appropriate quantity to use is the mean square fluctuation smoothed on a given scale R, which can be shown to be

$$\sigma^2(R) = 4\pi \int W^2(kR) P(k) k^2 \mathrm{d}k, \tag{6.18}$$

where W is the Fourier transform of the filter function used to smooth the density field in real space (e.g. a Gaussian). The value of $\sigma(R = 8h^{-1}\ \mathrm{Mpc})$ is often used for normalisation purposes; we call this σ_8 from now on. In the linear regime, the density field retains its initial Gaussian character: all moments of order $n > 2$ are either zero (odd) or can be expressed in terms of σ^2 (even).

As well as the amplitude, we also need to specify the shape of the *initial* fluctuation spectrum, which is taken to be imprinted on the universe at some arbitrarily early time. The inflationary scenario for the very early universe (Guth 1981; Linde 1982) often produces a power-law form

$$P(k) = Ak^n, \tag{6.19}$$

with many models having the so-called Harrison–Zel'dovich form with $n = 1$ (Harrison 1970; Zel'dovich 1972). Even if inflation is not the origin of density fluctuations, the form (6.19) is a useful phenomenological model for the fluctuation spectrum.

6.1.4 The transfer function

We have hitherto assumed that the effects of pressure and other astrophysical processes on the gravitational evolution of perturbations are negligible. In fact, depending on the form of any dark matter, and the parameters of the background cosmology, the growth of perturbations on particular length scales can be suppressed relative to the growth laws discussed above.

We need first to specify the fluctuation mode. In cosmology, the two relevant alternatives are *adiabatic* and *isocurvature*. The former involve coupled fluctuations in the matter and radiation component in such a way that the entropy does not vary spatially; the latter have zero net fluctuation in the energy density and involve entropy fluctuations. Adiabatic fluctuations are the generic prediction from inflation and, more importantly, no successful model of structure formation has yet been constructed based on isocurvature fluctuations.

In the classical analysis of Jeans (1902), pressure inhibits the growth of structure on scales smaller than the distance traversed by an acoustic wave during the free-fall collapse time of a perturbation. If there are collisionless particles of hot dark matter, they can travel rapidly through the background and this free streaming can damp away perturbations completely. Radiation and relativistic particles may also cause kinematic suppression of growth. The imperfect coupling of photons and baryons can also cause dissipation of perturbations in the baryonic component. The effects of these processes must be computed using a Boltzmann code to handle the transport of the matter distribution functions through the epoch from horizon entry to recombination. This kind of calculation is described in standard texts and will not be discussed here. The net effect of these processes, for the case of statistically homogeneous initial Gaussian fluctuations, is to change the shape of the original power-spectrum in a manner described by a simple function of wave-number – the transfer function $T(k)$ – which relates the processed power-spectrum $P(k)$ to its primordial form $P_0(k)$ via $P(k) = P_0(k) \times T^2(k)$. The results of full numerical calculations of all the physical processes we have discussed can be encoded in the transfer function of a particular model (Bardeen *et al.* 1986; Holtzmann 1989). For example, fast moving or 'hot' dark matter particles (HDM) erase structure on small scales by the free-streaming effects mentioned above so that $T(k) \to 0$ exponentially for large k; slow moving or 'cold' dark matter (CDM) does not suffer such strong dissipation, but there is a kinematic suppression of growth on small scales (to be more precise, on scales less than the horizon size at matter–radiation equality); significant small-scale power nevertheless survives in the latter case. These two alternatives thus furnish two very different scenarios for the late stages of structure formation: the 'top–down' picture exemplified by HDM first produces superclusters, which subsequently fragment to form galaxies; CDM is a 'bottom–up' model because small-scale structures form first and then merge to form larger ones.

The general picture that emerges is that, while the amplitude of each Fourier mode remains small, i.e. $\delta(\mathbf{k}) \ll 1$, linear theory applies. In this regime, each Fourier mode evolves independently and the power-spectrum therefore just scales as

$$P(k,t) = P(k,t_1)\frac{D_+^2(k,t)}{D_+^2(k,t_1)} = P_0(k)T^2(k)\frac{D_+^2(k,t)}{D_+^2(k,t_1)} \,. \qquad (6.20)$$

Even if $\delta > 1$ for large k, one can still apply linear theory on large scales by applying a filter in the manner represented by equation (6.18) to remove the non-linear modes. Generally, it is assumed that a scale evolves linearly as long as $\sigma^2(R) < 1$, though this is not always correct (e.g. Peebles 1980). For scales larger than the Jeans length, this means that the shape of the power-spectrum is preserved during linear evolution. Only when scales go non-linear does the shape of the spectrum begin to change. Of course, scales below the Jeans length do not grow relative to those above it, and, as mentioned above, in the popular models with non-baryonic dark matter, there are other effects suppressing the growth of fluctuations in the linear regime, so the shape of the spectrum is distorted even in the linear phase. For a detailed review of the behaviour of cosmological density perturbations, see Efstathiou (1990).

These considerations specify the shape of the fluctuation spectrum, but not its amplitude. Historically the approach taken to fix the amplitude has been empirical. For example, the rms fluctuation in galaxy counts has been related to the corresponding quantity calculated from the power-spectrum (6.18), using σ_8. However, this method of normalisation is problematic if there exists a bias (see below). The discovery of temperature fluctuations in the CMB (Smoot *et al.* 1992) has furnished an alternative procedure: assuming these fluctuations are due to scalar density perturbations, then their amplitude fixes the spectrum normalisation of large scales. The one ambiguity of this approach is that the temperature anisotropies may actually be caused by long-wavelength gravitational wave (tensor) perturbations instead, particularly if they are accompanied by a non-flat fluctuation spectrum (e.g. Lidsey & Coles 1992). Invoking such perturbations can allow one to reduce the spectrum normalisation relative to the case when they are entirely scalar in character. Measurements of CMB anisotropy and their implications for cosmology are discussed in more detail in Chapter 7.

6.1.5 Beyond linear theory?

The linearised equations of motion provide an excellent description of gravitational instability at very early times when density fluctuations are still small ($\delta \ll 1$). The linear regime of gravitational instability breaks down when δ becomes comparable to unity, marking the commencement of the *quasi-linear* (or weakly non-linear) regime. During this regime the density contrast may remain small ($\delta < 1$), but the phases of the Fourier components $\delta_{\mathbf{k}}$ become substantially different from their initial values resulting in the gradual development of a non-Gaussian distribution function if the primordial density field was Gaussian. In this regime the shape of the power-spectrum changes by virtue of a complicated cross-talk between different wave-modes. Analytic methods are available for this kind of problem (reviewed by, e.g., Sahni & Coles 1995), but the usual approach is to use N-body experiments for strongly non-linear analyses (e.g. Davis *et al.* 1985; Bertschinger & Gelb 1991).

Further into the non-linear regime, bound structures form. The baryonic content of these objects may then become important dynamically: hydrodynamical effects (e.g. shocks), star formation and heating and cooling of gas all come into play. The spatial distribution of galaxies may therefore be very different from the distribution of the (dark) matter, even on large scales. Attempts are only just being made to model some of these processes with cosmological hydrodynamics codes (e.g. Cen 1992), but it is some measure of the difficulty of understanding the formation of galaxies and clusters that most theoretical studies do not even attempt to take into account the detailed physics of galaxy formation. The usual approach is instead simply to assume that the point-like distribution of galaxies, galaxy clusters or whatever,

$$n(\mathbf{r}) = \sum_i \delta_D(\mathbf{r} - \mathbf{r}_i), \tag{6.21}$$

bears a simple functional relationship to the underlying $\delta(\mathbf{r})$. An assumption often invoked is that relative fluctuations in the object number counts and matter density fluctuations are proportional to each other, at least within sufficiently large volumes, according to the *linear biasing* prescription:

$$\frac{\delta n(\mathbf{r})}{\bar{n}} = b\,\frac{\delta\rho(\mathbf{r})}{\bar{\rho}}, \tag{6.22}$$

where b is what is usually called the biasing parameter. Alternatives, which are not equivalent, include the high-peak model (Kaiser 1984; Davis *et al.* 1985; Bardeen *et al.* 1986) and the various local bias models (Coles 1993). For a discussion of the possible physical mechanisms responsible for inflicting a bias on the galaxy distribution see Rees (1985), Dekel (1986), and Dekel & Rees (1987). Non-local biases are possible, but rather harder to construct plausibly (Babul & White 1991; Bower *et al.* 1993). If one is prepared to accept an *ansatz* of the form (6.22) then one can use linear theory on large scales to relate galaxy clustering statistics to those of the density fluctuations, e.g.

$$P_{\mathrm{gal}}(k) = b^2 P(k). \tag{6.23}$$

This approach is frequently adopted in practice.

6.1.6 The contenders

Before introducing the contenders for the eventual role of theory of structure formation, it is perhaps useful to summarise the previous rather lengthy sections on the relevant theoretical elements. In the following list of the various ingredients needed to specify a model of structure formation within this paradigm, notice that most of these ingredients involve at least one assumption that may well turn out not to be true:

(1) A background cosmology. This basically means a choice of Ω_0, H_0 and Λ, assuming we are prepared to stick with the Robertson–Walker metric (1.1) and the Einstein equations (1.3).
(2) An initial fluctuation spectrum. This is usually taken to be a power-law, but may not be. The most common choice is $n = 1$.
(3) A choice of fluctuation mode: usually adiabatic.
(4) A statistical distribution of fluctuations. This is often assumed to be Gaussian.
(5) The transfer function, which requires knowledge of the relevant proportions of 'hot', 'cold' and baryonic material as well as the number of relativistic particle species.
(6) The normalisation, either from COBE or by σ_8, assuming some value of b.
(7) A 'machine' for handling non-linear evolution, so that the distribution of galaxies and other structures can be predicted. This could be an N-body or hydrodynamical code, an approximated dynamical calculation or simply, with fingers crossed, linear theory.
(8) A prescription for relating fluctuations in mass to fluctuations in light, frequently the linear bias model (6.22).

Only after these components have been detailed can one compare the model against observations.

We shall not discuss the baryon-only models predominant in the 1970s and early 1980s, though these are historically important; discussions can be found in textbooks (e.g. Coles & Lucchin 1995). It is also worth mentioning that there is a radical alternative to the non-baryonic models in the form of the phenomenological baryonic isocurvature (PBI) model (e.g. Peebles 1987). It is, however, extremely doubtful whether this can match the observations and is regarded by many as a non-starter.

Historically speaking, the first model incorporating non-baryonic dark matter to be seriously considered was the hot dark matter (HDM) scenario, in which the universe is dominated by a massive neutrino with mass around 10–30 eV, as discussed in Chapter 2.

This scenario has fallen into disfavour because the copious free streaming it produces smooths the matter fluctuations on small scales and means that galaxies form very late. The favoured alternative for most of the 1980s was the cold dark matter (CDM) model in which the dark matter particles owing to their much higher mass (e.g. supersymmetric particles) or non-thermal distribution (e.g. axions) undergo negligible free streaming. A 'standard' CDM model (SCDM) then emerged in which the cosmological parameters were fixed at $\Omega_0 = 1$ and $h = 0.5$, the spectrum was of the Harrison–Zel'dovich form with $n = 1$ and a significant bias, $b = 1.5$ to 2.5, was required to fit the observations (Davis *et al.* 1985).

The SCDM model was ruled out by a combination of the COBE-inferred amplitude of primordial density fluctuations, galaxy clustering power-spectrum estimates on large scales, cluster abundances and small-scale velocity dispersions (e.g. Peacock & Dodds 1994); see below for further discussion. It seems the standard version of this theory simply has a transfer function with the wrong shape to accommodate all the available data with an $n = 1$ initial spectrum. Nevertheless, because CDM is such a successful first approximation and seems to have gone a long way to providing an answer to the puzzle of structure formation, the response of the community has not been to abandon it entirely, but to seek ways of relaxing the constituent assumptions in order to get a better agreement with observations. Various possibilities have been suggested.

If the total density is reduced to $\Omega_0 \simeq 0.2$, which is favoured by many of the arguments we present in this book, then the scale of the horizon at matter–radiation equivalence increases and much more large-scale clustering is generated. This is called the open cold dark matter model, or OCDM for short. A similar effect is produced if one is prepared to accept a very low value for the Hubble parameter, $h \simeq 0.3$ (Bartlett *et al.* 1995). Those unwilling to dispense with the inflationary predeliction for flat spatial sections have invoked $\Omega_0 = 0.2$ and a positive cosmological constant (Efstathiou, Sutherland & Maddox 1990) to ensure that $k = 0$; this can be called ΛCDM. This latter model is put under severe pressure by the statistics of multiply lensed quasar images (see Chapter 3) which seem to require $\Omega_0 > 0.3$ ($\Omega_\Lambda < 0.7$) if $k = 0$, but may at least escape from the formidable problems posed by the conflict between globular star cluster ages and the inverse of the Hubble parameter (Chapter 3).

Another possibility is to invoke non-flat initial fluctuation spectra, while keeping everything else in SCDM fixed. The resulting 'tilted' models, TCDM, usually have $n < 1$ power-law spectra for extra large-scale power and, perhaps, a significant fraction of tensor perturbations (Lidsey & Coles 1992). Models have also been constructed in which non-power-law behaviour is invoked to produce the required extra power: these are the broken scale-invariance (BSI) models (Gottlöber, Mücket and Starobinsky 1994).

Probably the most successful alternative to SCDM is to assume a mixture of hot and cold dark matter (CHDM), having perhaps $\Omega_{\rm hot} = 0.3$ for the fractional density contributed by the hot particles. For a fixed large-scale normalisation, adding a hot component has the effect of suppressing the power-spectrum amplitude at small wavelengths (Klypin *et al.* 1993). A variation on this theme would be to invoke a 'volatile' rather than 'hot' component of matter produced by the decay of a heavier particle (Pierpaoli *et al.* 1996). The non-thermal character of the decay products results in subtle differences in the shape of the transfer function in the (CVDM) model compared to the (CHDM) version.

This list illustrates most of the contenders for the title of 'standard model'. It has to be said that there is something unsatisfactory about simply adding extra parameters to the CDM model in the manner of the Ptolemaic epicycles. If, however, the existence of a neutrino with a mass of a few eV were to be established, the CHDM model at least would not be subject to this criticism. One should also be aware of the obvious point that Ω_0, Λ and H_0 are not simply free parameters: they should be fixed by other cosmological observations. In any case it is not being totally facetious to say that any model that fits all the observational data at any time is probably wrong because, at any given time, at least some of the data are likely be wrong.

6.2 Galaxy clustering

6.2.1 Redshift surveys

To give an idea of how much progress has been made in gathering information on galaxy clustering, it is helpful to compare Figure 6.1 with Figure 6.2. The CfA survey (de Lapparent *et al.* 1986) shown in Figure 6.1 was the 'state-of-the-art' in 1986, but contained only around 2000 galaxies with a maximum recession velocity of 15 000 km s^{-1}. The section of the Las Campanas survey displayed in Figure 6.2 shows around six times as many galaxies as Figure 6.1, and goes out to a velocity of 60 000 km s^{-1} (Shectman *et al.* 1996). Galaxies lying in a narrow strip on the sky are plotted on these pie diagrams with recession velocity representing distance using the Hubble law, equation (1.7). Data have also been accumulated for galaxies selected in the infrared using the IRAS satellite to complement the optical data shown in these figures. Impressive though these exercises in cosmography may be, what is really needed is a statistical measure of the clustering pattern in these surveys to be compared with similar statistical predictions of the theory. Measures of spatial clustering offer the simplest method of probing $P(k)$, assuming that these objects are related in some well-defined way to the mass distribution and this, through the transfer function, is one way of constraining cosmological parameters. At the present time, redshifts of around 10^5 galaxies are available. The next generation of redshift surveys, prominent among which are the Sloan Digital Sky Survey† of about one million galaxy redshifts and an Anglo-Australian collaboration using the two-degree field (2DF) multi-fibre spectroscope which can obtain 400 galaxy spectra in one go‡, will increase the number of redshifts by about two orders of magnitude over what is currently available. Data from these surveys will contain much more useful information than simply better estimates of $P(k)$ and will enable more sophisticated tests of the gravitational instability picture to be carried out.

† `http://www-sdss.fnal.gov:8000/`
‡ `http://msowww.anu.edu.au/~colless/2dF/`

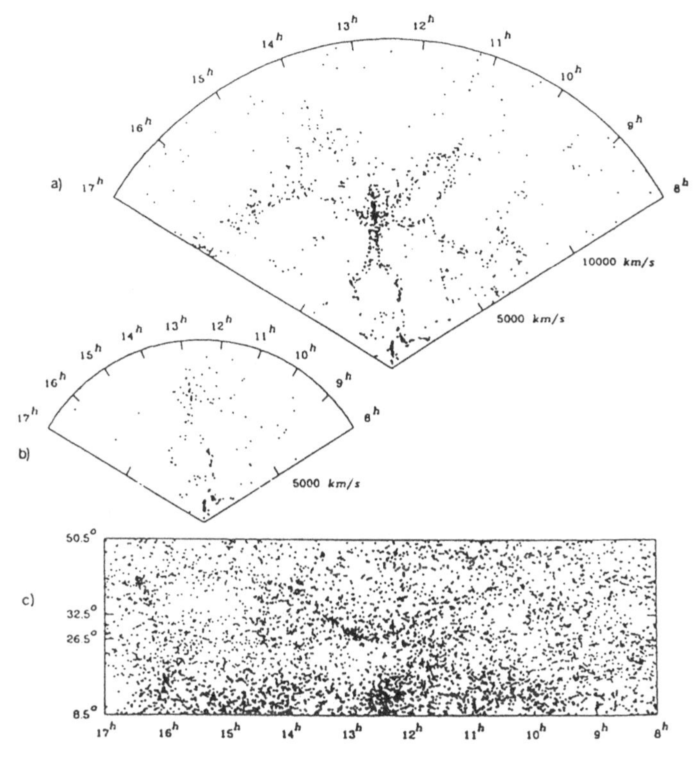

Fig. 6.1 A redshift survey from the 1980s: the famous 'CfA slice' (De Lapparent, Geller & Huchra 1986).

6.2.2 Galaxy and cluster correlations

The traditional tool (e.g. Peebles 1980) for studying galaxy clustering is the two-point correlation function, $\xi(r)$. This is defined by the probability δP of finding two objects (e.g. galaxies or clusters) in two disjoint volumes of space δV_1 and δV_2 separated by a distance r. Assuming statistical homogeneity and isotropy, so that this probability depends upon neither spatial position nor direction, we get

$$\delta P = n_v^2 [1 + \xi(r)] \delta V_1 \delta V_2, \qquad (6.24)$$

where n_v is the mean spatial number-density of objects.

The observed galaxy-galaxy correlation function (e.g. Davis & Peebles 1983) is approximated by a power law with index $\gamma \simeq 1.8$

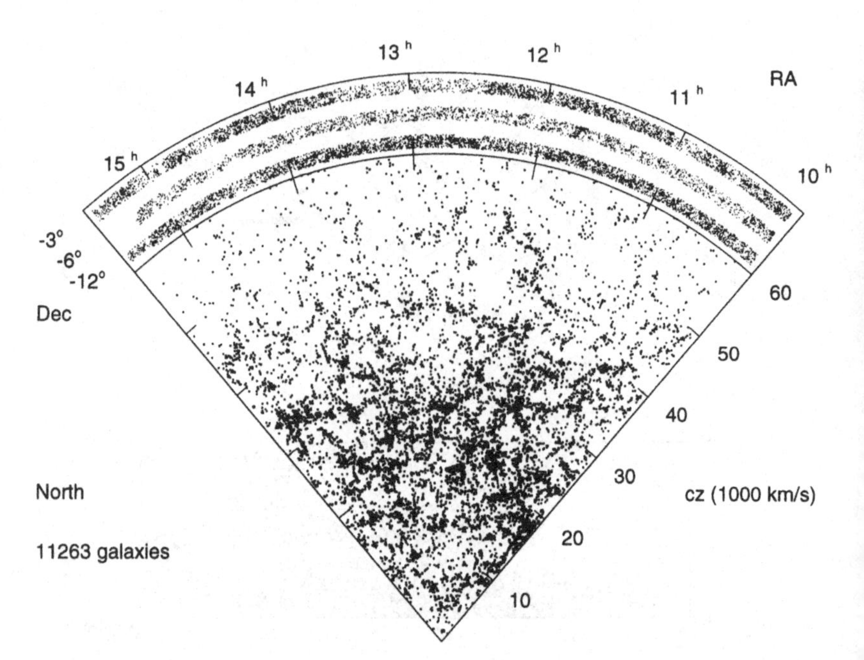

Fig. 6.2 A redshift survey from the 1990s: the new Las Campanas Redshift Survey (Shectman *et al.* 1996). Picture courtesy of Steve Shectman.

and correlation length $r_0 \simeq 5h^{-1}$ Mpc:

$$\xi(r) \simeq \left(\frac{r}{r_0}\right)^{-\gamma}. \tag{6.25}$$

On the other hand, the correlation function of rich Abell clusters has a roughly similar shape, but with a larger correlation length $r_0 \simeq (15-25)h^{-1}$ Mpc (Bahcall & Soneira 1983). Indeed, this latter observation provides much of the motivation for the high-peak bias model of Kaiser (1984) mentioned above, because clusters are clearly defined as high peaks of the cosmological density field.

On small scales, when $\xi(r) \gg 1$ is easily measurable, clustering is dominated by the (uncertain) form of the bias and by non-linear dynamical evolution so it is difficult to relate the results to anything primordial. On large scales, $\xi(r)$ is very difficult to measure, because of the restricted size of the samples available.

6.2.3 Angular correlations

One way to estimate $\xi(r)$ on large scales is to measure instead the angular correlation function $w(\theta)$, defined in the same way as $\xi(r)$ but in terms of angular separations rather than spatial ones, for galaxies observed in two-dimensional projection on the sky. One can then deproject the data, using the assumed separability of luminosity distributions and spatial distributions of galaxies, to recover $\xi(r)$. With automated plate-scanning devices, such as APM (Maddox *et al.* 1990) and COSMOS (Collins *et al.* 1992) it is feasible to measure $w(\theta)$ for samples of more than a million galaxies. The results of this kind of analysis have shown that $\xi(r)$ for optically selected galaxies and clusters is incompatible with the standard version of the CDM model (with $\Omega_0 = 1$ and $h = 0.5$). The results are compatible with a CDM model with $\Omega_0 = 0.2$, with or without a cosmological constant (Efstathiou, Sutherland & Maddox 1990; Bahcall & Cen 1992).

6.2.4 Fluctuations in counts

An alternative approach is to adopt a statistical measure which is better suited than $\xi(r)$ to measure very small correlations. One such method is to look at the fluctuations of galaxy counts in cells of equal volume. The expected variance of such cell counts can be expressed as an integral of the two-point correlation function over the cell volume, plus a correction for shot noise. This was the method favoured by the QDOT† surveys, in which a large redshift sample of galaxies selected in the infrared were analysed. Intriguingly, their results are also in agreement with the APM results over the range of scales they have in common (Efstathiou *et al.* 1990; Saunders *et al.* 1991), so they are also compatible with $\Omega_0 \simeq 0.2$.

† QDOT is the name given to a redshift survey of a sample of one in six galaxies identified by the IRAS infrared satellite: the acronym derives from Queen Mary, Durham, Oxford and Toronto.

6.2.5 Power-spectra

Alternatively, one can measure not $\xi(r)$ but its Fourier transform, namely the power-spectrum. This is especially useful because it is the power-spectrum which is predicted directly in cosmogonical models incorporating inflation and dark matter. Peacock (1991) and Peacock & Dodds (1994) have recently made compilations of power-spectra of different kinds of galaxy and cluster redshift samples and, for comparison, has included a deprojection of the APM $w(\theta)$. Within the (considerable) observational errors, and the uncertainty introduced by modelling of the bias, all the data lie roughly on the same curve; see Figure 6.3.

A coherent picture thus emerges in which galaxy clustering extends over larger scales than is expected in the standard CDM scenario. This excess can be quantified using the 'extra power' parameter, a rough measure of the shape of the power-spectrum which can be defined, for convenience, in terms of the ratio between the rms density fluctuations on scales of 25 and $8h^{-1}$ Mpc:

$$\Gamma = 0.5\left(\frac{4\sigma_{25}}{0.95\sigma_8}\right)^{0.3}; \tag{6.26}$$

$\Gamma \simeq \Omega_0 h = 0.5$ for SCDM, while the data suggest $\Gamma = 0.2$ to 0.3, which can be achieved by reducing Ω_0 or H_0 in the manner described above; the data are thus broadly consistent with a CDM-like power-spectrum but with $\Omega_0 \simeq 0.2$; see also Liddle *et al.* (1996). Further arguments on the qualitative consistency of this picture, taking into account the evolution of clustering with redshift, are given by Peacock (1996).

6.2.6 The abundances of objects

In addition to their spatial distribution, the number-densities of various classes of cosmic objects as a function of redshift can be used to constrain the shape of the power-spectrum. In particular, if objects are forming hierarchically there should be fewer objects of a given mass at high z and more objects with lower mass. According to the standard Press & Schechter (1974) approach, the number density of collapsed structures arising from Gaussian

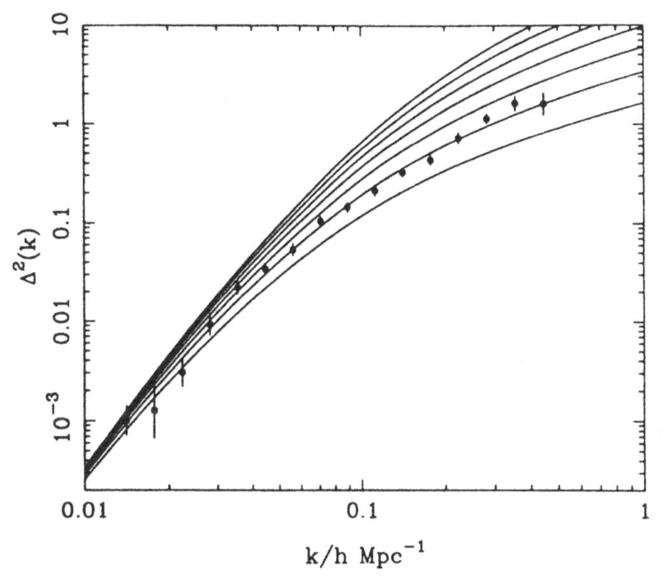

Fig. 6.3 Estimated power-spectra using a compilation of redshift survey data from various sources; from Peacock & Dodds (1994). The data have been 'corrected' for non-linear evolution, redshift-space distortion and bias. The box on the left-hand side of the figure shows the uncertainty inherent in the normalisation of the fluctuation spectrum on large scales to COBE. The dashed line shows the power-spectrum for the standard cold dark matter model (SCDM): it clearly overshoots the small-scale clustering data. The solid line is a model fit which roughly corresponds to the open cold dark matter (OCDM) model described in the text.

initial fluctuations and having masses larger than M is given by

$$N(>M) = \int_M^\infty n(M')\,\mathrm{d}M' . \tag{6.27}$$

Here $n(M)\,\mathrm{d}M$ is the number density of objects with mass in the range $[M, M+\mathrm{d}M]$ and is related to the power-spectrum according to

$$n(M)\,\mathrm{d}M = \frac{1}{\sqrt{2\pi}}\frac{\delta_c}{f}\int_R^\infty \frac{\eta_R}{\sigma_R}\exp\left(-\frac{\delta_c^2}{2\sigma_R^2}\right)\frac{\mathrm{d}R}{R^2}, \tag{6.28}$$

where

$$\eta_R = \frac{1}{2\pi^2 \sigma_R^2} \int k^4 \, P(k) \, \frac{dW^2(kR)}{d(kR)} \, \frac{\mathrm{d}k}{kR} \, . \qquad (6.29)$$

In the above expressions, we assume that the mass scale M is related to the length scale R according to $M = f \bar{\rho} R^3$, with f the 'form factor', which is specified by the shape of the filter W and $\bar{\rho}$ the average density. For the Gaussian window, which we assume here, it is $f = (2\pi)^{3/2}$, while $f = 4\pi/3$ for a top-hat window. The parameter δ_c is the critical density contrast, which represents the threshold value for a fluctuation to turn into an observable object, if evolved to the present time by linear theory. For a top-hat spherical collapse one has $\delta_c = 1.68$, but the inclusion of non-linear effects, as well as aspherical collapse, may lead to a lower value of δ_c. Values usually quoted are in the range $[1.4, 1.7]$.

In the context of rich galaxy clusters this constraint has been employed by White, Efstathiou & Frenk (1993) using X-ray data, and Biviano *et al.* (1993) using velocity dispersions to measure the cluster masses, giving rather different results due to systematic effects, related to assumptions used to connect X-ray temperature and dark-matter potential profiles, or to biases in estimating cluster masses from internal velocities under the virial assumption. A recent study of cluster abundances by Viana & Liddle (1996), taking into account these uncertainties, concludes that the data can be accomodated by CDM-like models for more-or-less any value of the density (within reason). The evolution of cluster numbers with redshift is a more sensitive probe of Ω_0 so that future studies of high-redshift clusters may yield more definitive results (Eke, Cole & Frenk 1996).

A further constraint on power-spectra comes from observations of high-redshift objects. The most reliable such constraint concerns the abundance of damped Lyman-α systems (DLAS); see §5.3.1. These are observed as wide absorption troughs in quasar spectra, owing to a high HI column density ($\geq 10^{20}$ cm^{-2}). The fact that at $z \sim 3$ the fractional density of HI gas associated with DLAS is comparable to that contributed by visible matter in nearby galaxies suggests that DLAS actually trace a population of collapsed protogalactic objects; see Wolfe (1993) for a comprehensive review. Lanzetta *et al.* (1995) and Wolfe *et al.* (1995) presented data on DLAS up to the redshift $z \simeq 3.5$, while the recent compilation by Storrie-Lombardi *et al.* (1995) pushed this limit to $z \simeq 4.25$. Based on these data, the latter authors claimed the first detection of a turnover in the fractional density, Ω_g, of neutral gas belonging to the absorbing systems at high redshift. In order to connect model predictions from linear theory with observations, let

$$\Omega_{\rm coll}(M,z) = {\rm erfc}\left(\frac{\delta_c}{\sqrt{2}\sigma(M,z)}\right) \qquad (6.30)$$

be defined as the fractional density contributed at the redshift z by collapsed structures of mass larger than M. Accordingly, $\Omega_g = \alpha_g \Omega_b \Omega_{\rm coll}$, where α_g is the fraction of HI gas which is involved in the absorbers.

Several authors have suggested DLAS as a powerful test for DM models, based on a linear theory approach. Kauffman & Charlot (1994), Mo & Miralda-Escudé (1994) and Ma & Bertschinger (1994) have concluded that the standard CHDM scenario with $\Omega_\nu = 0.3$ is not able to generate enough collapsed structures at $z > 3$, owing to the lack of power on galactic scales. However, either lowering Ω_ν to about 0.2 (Klypin *et al.* 1995) or 'blueing' the primordial spectrum so that $n \simeq 1.2$ (Borgani *et al.* 1996) keeps CHDM in better agreement with the data. Katz *et al.* (1996), using numerical simulations of the formation of DLAS, found that even the CDM model with a normalisation as low as $\sigma_8 = 0.7$ satisfies the DLAS constraint. Again, the results are not definitive, but future observational programmes may strengthen these constraints.

6.3 Dipole analysis

It has been known for many years that the CMB radiation possesses a dipole anisotropy corresponding to the motion of the Earth through a frame in which the radiation is isotropic (Lubin, Epstein & Smoot 1983). After subtracting the Earth's motion around the Sun, the Sun's motion around the Galactic centre and the velocity of our Galaxy with respect to the centroid of the Local Group, this dipole anisotropy tells us the speed and direction of the Local Group through the cosmic reference frame. The result is a velocity of 600 km s^{-1} in the direction of Hydra–Centaurus ($l = 268°$, $b = 27°$). This can be used to estimate Ω_0.

6.3.1 The basic idea

In the gravitational instability picture the Local Group velocity can be explained as being due to the net gravitational pull on the Local Group owing to the inhomogeneous distribution of matter around it. In fact the net gravitational acceleration is just

$$\mathbf{g} = G \int \frac{\rho(\mathbf{r})\mathbf{r}}{r^2} \mathrm{d}^3 r, \tag{6.31}$$

where the integral should formally be taken to infinity; this equation will appear again in a slightly different form in (6.34). In the linear theory of gravitational instability, this gravitational acceleration is just proportional to, and in the same direction as, the net velocity. From equation (6.12) the constant of proportionality depends on $f \simeq \Omega_0^{0.6}$. If one can estimate ρ from a sufficiently large sample of galaxies then one can in principle determine Ω_0. Of course, the ubiquitous bias factor (6.22) intrudes again, so that one can actually only determine $\Omega_0^{0.6}/b$, and even that only as long as b is constant. For this reason, techniques using velocities generally yield constraints on the parameter

$$\beta \equiv \frac{\Omega_0^{0.6}}{b}. \tag{6.32}$$

The technique is simple. Suppose we have a sample of galaxies with some well-defined selection criterion so that the selection function – the probability that a galaxy at distance r from the observer is included in the catalogue – has some known form $\phi(r)$. Then $\mathbf{g}$ can be approximated by

$$\mathbf{g} = \frac{4\pi G}{3}\mathbf{D} = G \sum_i \frac{1}{\phi(r_i)} \frac{\mathbf{r_i}}{r_i^2}, \tag{6.33}$$

where the $\mathbf{r_i}$ are the galaxy positions and the sum is taken over all the galaxies in the sample. The dipole vector $\mathbf{D}$ can be computed from the catalogue and, as long as it is aligned with the CMB dipole anisotropy, one can estimate β. It must be emphasised that this method measures only the inhomogeneous component of the gravitational field: it will not detect a mass component that is uniform on a scale larger than can be probed by available samples.

6.3.2 The IRAS dipole

This technique has been very popular with cosmologists over the last few years, mainly because the various IRAS galaxy catalogues are very suitable for this type of analysis (Strauss & Davis 1988a,b; Rowan-Robinson *et al.* 1990). There are, however, a number of difficulties which need to be resolved before cosmological dipoles can seriously be said to yield an accurate determination of $\Omega_0^{0.6}$, or even β.

First, and probably most importantly, is the problem of convergence. Suppose one has a catalogue that samples a small sphere around the Local Group but that this sphere is is itself moving in the same direction (or similar) as the Local Group. For this to happen, the universe must be significantly inhomogeneous on scales larger than the catalogue can probe. In this circumstance, the actual velocity explained by the dipole of the catalogue is not the whole CMB dipole velocity but only a part of it. It follows then that one would overestimate the β factor by comparing too large a velocity with the observed $\mathbf{D}$. One must be absolutely sure, therefore, that the sample is deep enough to sample all contributions to the Local Group motion. The history of this kind of analysis is littered with examples of failures to realise this simple fact. The early analyses of IRAS galaxy samples yielded high values of $\Omega_0^{0.6}$, but it was clear that the dipole kept growing until the number-density of galaxies in the sample was very small. Strauss & Davis (1988a,b) claimed that the IRAS dipole converged within $60h^{-1}$ Mpc of the Local Group. A deeper sample, the QDOT redshift survey, analysed by Rowan-Robinson *et al.* (1990), demonstrates that this is not true but that there are significant contributions to the dipole at least as far as $100h^{-1}$ Mpc. It has been claimed that the QDOT dipole has clearly converged by the limit of this catalogue. In our view, there is absolutely no evidence to support this claim. The trouble is that the selection function for these galaxies is so small at the relevant scales that there is no way it would register any inhomogeneities above the statistical noise. Indeed, the QDOT dipole behaves at $100h^{-1}$ in an exactly comparable manner to the way the Strauss–Davis dipole behaves at $60h^{-1}$ Mpc.

6.3.3 Rich clusters of galaxies

So can one resolve the problem of the selection function and see if there is indeed a significant contribution to **D** from large scales? One clearly needs an object to trace the mass which does not have as steep a selection function as IRAS galaxies. Rich clusters seem to fit the bill nicely: suitably chosen samples have selection functions which fall relatively slowly even beyond the QDOT depth. The problem with these samples is that they are very sparse, so the sampling of the density field is not good, and clusters are probably biased tracers of the density field (Kaiser 1984). Analysis of cluster samples (Scaramella, Vettolani & Zamorani 1991; Plionis & Valdarnini 1991) shows very clearly that there is a large contribution to the dipole from scales as large as $150h^{-1}$ Mpc. There is even circumstantial evidence for similar behaviour in the QDOT dipole itself; see Plionis, Coles & Catelan (1992).

In our review article for *Nature* (Coles & Ellis 1994), we stressed the behaviour of the cluster dipole as being evidence in support of a low value of Ω_0 in the range 0.3–0.4, based on these analyses. Subsequent study of this problem has, however, caused us to re-think the strength of the constraints provided by the cluster dipole. In particular, detailed simulation work by Tini Brunozzi *et al.* (1995) has demonstrated that the cluster dipole is a very unreliable indicator of Ω_0. What was done in this work was to simulate the observational samples of clusters as realistically as possible, in two models, one with $\Omega_0 = 1$ and the other with $\Omega_0 = 0.2$. An ensemble of observers in each set of simulations was constructed and the observational procedure followed to yield an estimate of Ω_0 for each observer and hence a distribution of estimated values of Ω_0. Although the mean value of Ω_0 corresponded to the expectation value in the model, i.e. either $\Omega_0 = 0.2$ or $\Omega_0 = 1$, the distributions around these values were large and highly skewed, particularly in the high-density model. Indeed, in the latter case, the most frequent value of Ω_0 was around 0.2; see Figure 6.4. There thus appears to be a kind of conspiracy (induced by bias and sampling effects) in the behaviour of the cluster dipole that could fool us into believing that the density were low even if it were high. We cannot therefore claim, as we did in Coles & Ellis (1994), that the cluster dipole provides strong evidence for a

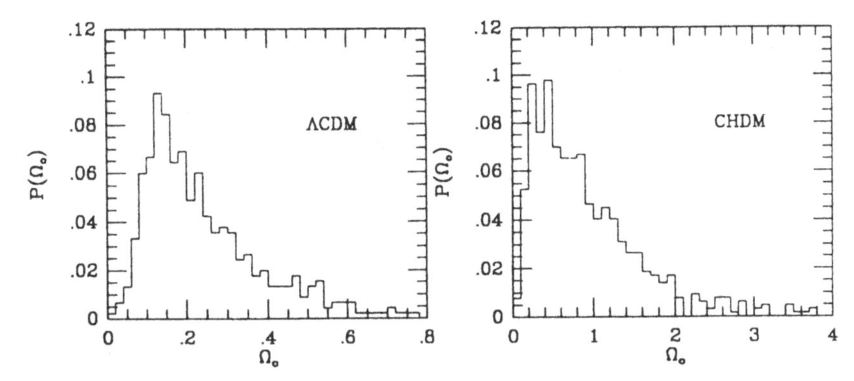

Fig. 6.4 Distribution of recovered values of Ω_0 for simulated cluster dipoles. From Tini Brunozzi *et al.* (1995).

low-density universe. Other properties of cluster motions are also now known to be compatible with the high-density case (Branchini, Plionis & Sciama 1996; Moscardini *et al.* 1996). The most reasonable conclusion to draw from this study is clearly that the cluster dipole is not a very good way to estimate Ω_0.

Nevertheless detailed simulations do confirm that a sample depth of at least $150h^{-1}$ Mpc is required for convergence of the dipole of the matter distribution (Kolokotronis *et al.* 1996) so the point we made above about the possibility of non-convergence at the depth of the QDOT sample suggests that caution should be exercised in the interpretation of the results. Note also that measurements of the dipole of the distribution of optical galaxies (e.g. Lahav 1987; Lahav, Rowan-Robinson & Lynden-Bell 1988; Lynden-Bell, Lahav & Burstein 1989; Kaiser & Lahav 1989; Lahav, Kaiser & Hoffman 1990; Hudson 1993) also lead to a low value of Ω_0, or at least β. Unless one has good reason to suppose that IRAS galaxies should be better tracers of the mass distribution than optical galaxies, this is further reason to doubt the consistency of the whole picture.

Incidentally, the fact that a considerable part of the cluster dipole comes from a distance of $\sim 140h^{-1}$ Mpc, roughly the same distance as the famous 'Shapley concentration' of galaxy clusters, has suggested to many that this is what is actually causing the pull on the Local Group. A careful look at the cluster data demonstrates that this cannot be true. The net dipole contributed by a shell in distance space at this scale would point in the direction of Shapley if this were the case. It does not. It points in the CMB dipole direction. To accept that there is such a contribution to the global dipole requires one to accept a very large scale inhomogeneity of which the Shapley concentration is only a part.

6.3.4 Other sources of error

We only summarise the other possible sources of error in the dipole analysis. First, one must ensure that the selection function is known very accurately, especially at large r. This essentially means knowing the luminosity function very accurately, particularly for the brightest objects (the ones that will be seen at great distances). There is also the problem that galaxies – in particular, IRAS galaxies – may be evolving with time, so the luminosity function for distance galaxies would be different from that of nearby ones. There is also the problem of bias. We have assumed a linear bias throughout the above discussion (as, indeed, has every analysis of cosmological dipoles published to date). Actually, there is little motivation for expecting this to be the case. The ramifications of non-linear and/or non-local biases have yet to be worked out in any detail. Furthermore, the direction of the dipole appears to be not too far away from the zone of avoidance caused by extinction in the plane of our own Galaxy. It is therefore possible that a large part of the mass responsible for the Local Group motion may be associated with objects in the Galactic plane which are difficult to include in surveys for observational reasons. And last, but by no means least, is the effect of redshift space distortions. At the sample depths required to say anything reasonable about large-scale structure, it is not practicable to obtain distances for all the objects, so one usually uses redshifts to estimate distances. At large r, one might expect this to be a good approximation. But, as Kaiser (1987) has shown, working in redshift space rather than distance space introduces rather alarming distortions into the analysis. One can illustrate some of the problems with the following toy example. Suppose an observer sits in a rocket and flies through a uniform distribution of galaxies. If he looks at the distribution in redshift space, even if the galaxies have no peculiar motions, he will actually see a dipole anisotropy caused by his motion. He may, if he is unwise, thus determine Ω_0 from his own velocity and this observed dipole: the answer would, of course, be entirely spurious and nothing whatsoever to do with the mean density of the universe.

The 'rocket effect' is not the only problem caused by redshift space distortions: such is the complexity of the problems caused by this phenomenon that even the *sign*, never mind the magnitude, cannot be calculated *a priori* for the QDOT data (Plionis, Coles & Catelan 1992).

6.4 Galaxy peculiar velocities

We have already mentioned the main problem of relying on a statistical analysis of the spatial distribution of cosmic objects to test theories: the bias. In an extreme case of bias one might imagine galaxies to be just 'painted on' to the background distribution in some arbitrary way having no regard to the distribution of mass. Ideally, one would wish to have some way of studying all the mass, not just that part of it which happens to light up.

One way to do this is to look at not just the spatial positions of galaxies, but also at their peculiar velocities. There are many ways to use the properties of peculiar motions in the estimation of Ω_0 and these have recently been the subject of two major technical review articles (Dekel 1994; Strauss & Willick 1995), so we shall not go into elaborate detail here, preferring to focus on perhaps the most robust methods and on the key results obtained. We shall also mainly concentrate on large-scale motions, as these are probably less clouded by the effects of non-linearity and bias.

6.4.1 Bulk flows and streaming motions

The most obvious way of exploiting galaxy motions is by trying to map out the flow pattern by directly measuring the peculiar velocity of each galaxy. The problem is that one must measure the galaxy distance using some indicator other than its redshift (e.g. Burstein 1990). The usual method is to use the Tully–Fisher relation for spiral galaxies and the D_n–σ relation for elliptical galaxies to obtain distance estimates; see Strauss & Willick (1995) for details. Once one knows the distance, one can subtract the contribution of a pure Hubble expansion at that distance to yield the galaxy's peculiar velocity. If peculiar velocities are supposed to be generated by inhomogeneities in the local distribution of material one can imagine using them to estimate the mass involved, in much the same way as one estimates the mass of the Sun from the Earth's orbital velocity. The trouble is that individual galaxy distance estimates are likely to be uncertain by at least 15% and the velocity field traced by galaxies is consequently very noisy. The usual way to use the peculiar velocity field is thus to measure *bulk flows* (sometimes called *streaming motions*), which represent the net motion of a large region, usually a sphere centred on the observer, in some direction relative to the pure Hubble expansion. The idea is that the spatial averaging smooths away the effect of distance errors. For instance, Bertschinger *et al.* (1990) report that a sphere of radius $40h^{-1}$ Mpc is executing a bulk flow of some 388 ± 67 km s^{-1} relative to the cosmological rest frame; a larger sphere of radius $60h^{-1}$ Mpc is moving at 327 ± 84 km s^{-1} (Bertschinger *et al.* 1990). How can one relate this type of measurement to theory?

In the linear theory of gravitational instability, one can relate the velocity of a matter at particular spatial position to the net gravitational acceleration at that point:

$$\mathbf{v}(\mathbf{x}) = H_0 \frac{f(\Omega_0)}{4\pi} \int \frac{\delta(\mathbf{x}')(\mathbf{x} - \mathbf{x}')}{|\mathbf{x} - \mathbf{x}'|^3} \mathrm{d}^3\mathbf{x}', \tag{6.34}$$

where f is defined by equation (6.8). There are various ways to exploit (6.34), which is merely equation (6.31) in a different guise, in order to obtain information about the density field δ on large scales (where linear theory is assumed to be applicable). If one can measure the rms bulk flow on some scale of averaging R, then this will be relating both to the power-spectrum $P(k)$ and $\Omega_0^{0.6}$. If one can estimate $P(k)$ then, in principle, one can determine Ω_0. If one estimates the power-spectrum from clustering data then there is an uncertainty of a factor b in the comparison, so one ends up estimating β. Measurements of the bulk flow in a sphere constitutes a measurement at a single location: there is a large uncertainty going from these to the global rms motion, which is what theory provides (in the form of an integral over the power-spectrum).

Recall from §6.1 that one can smooth the density perturbation field to define a mass variance in the manner of equation (6.18). One can do the same thing with the velocity field. If the density field is Gaussian, then so will be each component of $\mathbf{v}$. The magnitude of the averaged peculiar velocity, v, will therefore possess a Maxwellian distribution:

$$P(v)\mathrm{d}v = \sqrt{\frac{54}{\pi}}\left(\frac{v}{\sigma_v}\right)^2 \exp\left[-\frac{3}{2}\left(\frac{v}{\sigma_v}\right)^2\right]\frac{\mathrm{d}v}{\sigma_v} . \qquad (6.35)$$

In these equations $\mathbf{v}$ represents the smoothed velocity field, i.e.

$$\mathbf{v} = \mathbf{v}(\mathbf{x}; R) = \frac{1}{(2\pi)^3}\int \tilde{\mathbf{v}}(\mathbf{k})W_v(\mathbf{k}; R)\exp(-i\mathbf{k}\cdot\mathbf{x})\mathrm{d}\mathbf{k}, \qquad (6.36)$$

where $W_v(\mathbf{k}; R)$ is a suitable window function with a characteristic scale R; $\tilde{\mathbf{v}}(\mathbf{k})$ is the Fourier transform on the unsmoothed velocity field $\mathbf{v}(\mathbf{x}; 0)$. From equation (6.33) we find that

$$\sigma_v^2(R) = \frac{(H_0 f)^2}{2\pi^2}\int_0^\infty P(k)W_v^2(kR)\mathrm{d}k, \qquad (6.37)$$

by analogy with equation (6.18). In this equation (6.37), σ_v is the rms value of $v(\mathbf{x}; R)$, where the mean is taken over all spatial positions $\mathbf{x}$. Clearly the global mean value of $\mathbf{v}(\mathbf{x}, R)$ must be zero in a homogeneous and isotropic universe. It is a consequence of equation (6.35) that there is a 90% probability of finding a measured velocity satisfying the constraint

$$\frac{\sigma_v}{3} \leq v \leq 1.6\sigma_v. \qquad (6.38)$$

The window function W_v must be chosen to model the way the sample is constructed. This is not completely straightforward because the observational selection criteria are not always well controlled and the results are quite sensitive to the shape of the window function.

Because the integral in equation (6.37) is weighted towards lower k than the definition of σ_M^2 given by equation (6.18), which has an extra factor of k^2, bulk flows are potentially useful for probing the linear regime of $P(k)$ beyond what can be reached using properties of the spatial clustering of galaxies. The problem is that one typically has one measurement of the bulk flow on a scale R and this does not provide a strong constraint on σ_v or $P(k)$, as is obvious from equation (6.38): if a theory predicts an rms bulk flow of 300 km s^{-1} on some scale, then a randomly selected sphere on that scale can have a velocity between 100 and 480 km s^{-1} with 90% probability, an allowed error range of a factor of almost five. Until much more data become available, therefore, such measurements can only be used as a consistency check on models and do not strongly discriminate between them. Velocities can, however, place constraints on the possible existence of bias since σ_v is simply proportional to b (in the linear bias model). For example, the standard CDM model predicts a bulk flow on the scale of $40h^{-1}$ Mpc of around 180 km s^{-1} if $b = 1$. This reduces to 72 km s^{-1} if $b = 2.5$, which was, at one time, the favoured value. The observation of a velocity of 388 km s^{-1} on this scale is clearly incompatible with CDM with this level of bias; it is, however, compatible with a $b = 1$ CDM model. Clearly the presence of the factor f in equation (6.34) means that high values of v tend to favour higher values of f and therefore higher values of Ω.

One should also remember that this kind of analysis is predicated on there being large-scale homogeneity of the matter distribution. The bulk velocity in larger and larger volumes should get smaller and smaller until, on sufficiently large scales, the volume considered is at rest with respect to the frame in which the cosmic microwave background is isotropic. Couched in the language of the previous section, this means that the matter dipole should converge accurately with the CMB dipole on large scales; Ellis & Baldwin (1984) propose a way of testing the consistency of this supposition that has yet to be carried out. A somewhat alarming study by Lauer & Postman (1994) has estimated the bulk velocity excecuted by a shell of galaxy clusters of radius $\simeq$ 15 000 km s^{-1}. They find a large velocity $\simeq$ 600 km s^{-1} which is *not* aligned with the CMB dipole. This phenomenon is very difficult to accommodate in most theoretical models (e.g. Moscardini *et al.* 1996) and, if it is verified, will pose great difficulties for cosmology! At the moment, however, the consensus amongst most cosmologists is that some as yet unknown systematic error in the distance indicator used – a photometric indicator based on the properties of brightest galaxies in clusters – is responsible for this result. If this turns out not to be the case, then it will be back to the drawing board for all of us.

6.4.2 Velocity-density reconstruction

A more sophisticated approach to the use of velocity information is provided by a relatively new and extremely ingenious approach that has been developed known as POTENT (Bertschinger & Dekel 1989; Dekel, Bertschinger & Faber 1990; Dekel *et al.* 1993). This procedure exploits the fact that, in the linear theory of gravitational instability, the velocity field is curl-free and can therefore be expressed as the gradient of a potential as in equation (6.12). This velocity potential turns out to be simply proportional to the linear theory value of the gravitational potential; cf. equation (6.34).

Because the velocity field is the gradient of a potential Φ_v, one can use the purely radial motions, v_r, revealed by redshift and distance information to map Φ_v in three dimensions:

$$\Delta\Phi_v(r,\theta,\phi) = -\int_0^r v_r(r',\theta,\phi)dr'. \tag{6.39}$$

It is not required that paths of integration be radial, but they are in practice easier to deal with.

Once the potential has been mapped, one can solve for the density field using the Poisson equation. This means therefore that one can compare the density field as reconstructed from the velocities with the density field measured directly from the counts of galaxies. This, in principle, enables one to determine directly the level of bias present in the data. The only other parameter involved in the relation between $\mathbf{v}$ and δ is then f, which, in turn, is a simple function of Ω. POTENT holds out the prospect, therefore, of supplying a measurement of Ω which is independent of b, unlike those discussed above. It seems to be the case that POTENT results are consistent with $\Omega_0 \simeq 1$, although the precise results obtained vary considerably with the way one does the comparison. Attempts simply to measure β do, however, seem to yield $\beta \sim 1$, which is consistent with $\Omega_0 \simeq 1$ if the galaxies are unbiassed. See Strauss & Willick (1993, Table 3) for a compilation of results obtained.

At this point, however, it is worth mentioning some of the possible problems with the POTENT analysis. As always, one is of course limited by the quality and quantity of the velocity data available. The distance errors, together with the relative sparseness of the data sets available, combine to produce a velocity field

$\mathbf{v}$ which is quite noisy. This necessitates a considerable amount of smoothing, which is also needed to suppress small-scale non-linear contributions to the velocity field. The smoothed field is then interpolated to produce a continuous field defined on a grid. The favoured smoothing is of the form

$$v_r(\mathbf{r}) = \sum_i W_i(\mathbf{r}) v_{r,i}, \tag{6.40}$$

where i labels the individual objects whose radial velocities, $v_{r,i}$, have been estimated, and the weighting function $W_i(\mathbf{r})$ is taken to be

$$W_i(\mathbf{r}) \propto n_i^{-1} \sigma_i^{-2} \exp\left(-\frac{|\mathbf{r} - \mathbf{r_i}|^2}{2R_S^2}\right); \tag{6.41}$$

n_i is the local number density of objects, σ_i is the estimated standard error of the distance to the ith object and R_S is a Gaussian smoothing radius, typically of the order of $12h^{-1}$ Mpc. If one uses clusters instead of individual galaxies, then σ_i can be reduced by a factor equal to the square root of the number of objects in the cluster, assuming the errors are random. One effect of the heavy smoothing is that the volume probed by these studies consequently contains only a few independent smoothing volumes and the statistical significance of any reconstruction is bound to be poor.

Notice that the potential field one recovers then has to be differentiated to produce the density field, which will again exaggerate the level of noise. (It is possible to improve on the linear solution to the Poisson equation by using the Zel'dovich approximation to calculate the density perturbation δ from the velocity potential.) The scale of the noise problem can be gauged from the fact that a 20% distance error is of the same order as the typical peculiar velocity for distances beyond $30h^{-1}$ Mpc.

Apart from the problem of noise, there are also other sources of uncertainty in the applicability of this method. In any redshift survey one has to be careful to control selection biases, such as the Malmquist bias, which can enter in a complicated and inhomogeneous way into this analysis. While much effort has indeed been expended in this direction, it is still not clear whether an inhomogeneous Malmquist bias is correctly accounted for in the POTENT analysis (Newsam *et al.* 1993). One also needs to believe that the distance indicators used are accurate. Most workers in this field claim that their distance indicators are accurate to, say, 10–20%. However, if the errors are not completely random, i.e. there is a systematic component which actually depends on the local density, then the results of this type of analysis can be seriously affected. In this case the systematic error in $\mathbf{v}$ correlates with density in a similar way to that expected if the velocities were generated dynamically from density fluctuations. For instance, it has been suggested that there is such a systematic error in the commonly used D_n–σ indicator for elliptical galaxies (Guzman & Lucey 1993). What happens is that old stellar populations produce a different response in the distance indicator compared to young ones. Since older galaxies formed earlier and in higher density environments, the upshot is exactly the sort of systematic effect that is so dangerous to methods like POTENT. Guzman & Lucey (1993) found that applying a corrected distance indicator to their sample of elliptical galaxies essentially eliminated *all* the observed peculiar motions, which means that the motions derived using the uncorrected indicator were completely spurious. Whether this type of error is sufficiently widespread to affect all peculiar motion studies is unclear but it suggests one should regard these results with some scepticism.

The possible existence of this kind of spurious peculiar motion is a good reason also to be cautious in interpreting the agreement between 'recovered' density fields and the observed spatial distribution of galaxies. It has been argued that the apparent agreement between these two puts strong constraints on the possible level of bias between galaxies and mass. However, if the velocity field is caused by a spurious correlation of the behaviour of the distance indicator with environment then one would also see a good agreement between the recovered field and the galaxy distribution. In the presence of this uncertainty, it cannot be said that POTENT places any strong constraints on either the bias or Ω_0. However good the method, garbage in will produce garbage out.

Additionally, the sparseness with which the local peculiar velocity field is sampled by these studies requires a prodigious smoothing of the data to yield any results in the face of shot noise. With this smoothing, one cannot resolve anything other than the gross features of the flow. The volume probed by these studies consequently contains only a few independent smoothing volumes, and the statistical significance of any reconstruction is bound to be poor. Moreover, the single-volume measurements suffer from the effect of cosmic variance. A measurement of the density in our locality does not necessarily constitute an accurate determination of the mean value if the distribution is highly inhomogeneous.

6.4.3 Redshift-space distortions

The methods we have discussed in the preceding sections require one to know peculiar motions for a sample of galaxies. There is an alternative approach, which does not need such information, and which may consequently be more reliable. This relies on the fact that peculiar motions affect radial distances and not tangential ones. The distribution of galaxies in 'redshift-space' is therefore a distorted representation of their distribution in real space. For example, dense clusters appear elongated along the line-of-sight because of the large radial velocity component of the peculiar velocities: the angular coordinates of galaxies are not affected. This effect is colloquially known as the 'Fingers of God' and it can be seen, for example, in Figure 6.1. Similarly, the correlation functions and power-spectra of galaxies should be expected to show a characteristic distortion when they are viewed in redshift space rather than in real space. This is the case even if the real space distribution of matter is statistically homogeneous and isotropic. The line-of-sight anisotropy introduced into statistical measures like the correlation function and power-spectrum can be used to estimate the magnitude of the radial component of the typical galaxy peculiar velocity. The typical galaxy relative peculiar velocities obtained by such methods are around 300 km s^{-1}.

Let us illustrate these ideas by considering the effect of these distortions upon the two-point correlation function of galaxies. The conventional way to describe this phenomenon is to define coordinates as follows. Consider a pair of galaxies with measured redshifts corresponding to velocities $\mathbf{v}_1$ and $\mathbf{v}_2$. The separation in redshift space is then just

$$\mathbf{s} = \mathbf{v}_1 - \mathbf{v}_2; \tag{6.42}$$

an observer's line of sight is defined by

$$\mathbf{l} = \frac{\mathbf{v}_1 + \mathbf{v}_2}{2} \tag{6.43}$$

and the separations parallel and perpendicular to this direction are then just

$$\pi = \frac{\mathbf{s} \cdot \mathbf{l}}{|\mathbf{l}|} \tag{6.44}$$

and

$$r_p = \sqrt{\mathbf{s} \cdot \mathbf{s} - \pi^2}, \tag{6.45}$$

respectively. When the correlation function is plotted in the π–r_p plane, redshift distortions produce two effects: a stretching of the contours of ξ along the π axis on small scales (less than a few Mpc) owing to non-linear pairwise velocities, and compression along the π axis on larger scales owing to bulk (linear) motions; we discuss the latter effect below.

Linear theory cannot be used to calculate the first of these contributions, so one has to use explicitly non-linear methods. The usual approach is to use the equation

$$\frac{\partial \xi}{\partial t} = \frac{1}{ax^2} \frac{\partial}{\partial x}[x^2(1+\xi)v_{12}], \tag{6.46}$$

which expresses the conservation of particle pairs; x is a comoving coordinate and $v_{12} = |\mathbf{s}|$. Equation (6.46) is actually the first of an infinite set of equations known as the BBGKY hierarchy. To close the hierarchy one needs to make an assumption about higher moments, see Peebles (1980), leading to the *cosmic virial theorem*:

$$\langle v_{12}^2(r) \rangle \simeq C_\gamma H_0^2 Q \Omega r_{0g}^\gamma r^{2-\gamma}, \tag{6.47}$$

where $C_\gamma \simeq 23.8$ if $\gamma = 1.8$ and Q is a constant of the order of unity. Assuming that the radial anisotropy in $\xi(r_p, \pi)$ is due to the velocities v_{12} then one can, in principle, determine an estimate of Ω from the small-scale anisotropy. Generally speaking, results obtained yield Ω_0 in the range 0.1–0.3 (Davis & Peebles 1983; Peebles 1984). Notice, however, that there is an implicit assumption that the galaxy correlation function and the mass covariance function are identical, so this estimate will depend upon b in a non-trivial way. Indeed, it was the failure of clustering models – specifically the CDM model with $\Omega_0 = 1$ – to reproduce simultaneously the correct clustering amplitude $\xi(r)$ and the observed peculiar velocity dispersions that was the stimulus to introduce a bias factor into the calculations (Davis *et al.* 1985). If galaxies are more clustered than the total amount of gravitating matter, then the analysis of redshift distortions leads to an incorrect determination of Ω_0.

Recent advances in the accumulation of galaxy redshifts have made it possible to attempt analyses of redshift space distortions on large scales (e.g. Kaiser *et al.* 1991; Hamilton 1992, 1993; Fisher, Sharf & Lahav 1994; Fisher *et al.* 1994; Cole, Fisher & Weinberg 1995; Heavens & Taylor 1995). On large scales one sees a qualitatively different effect from the small-scale redshift distortions: a large-scale overdensity will tend to be collapsing in real space. Matter will therefore be moving towards a cluster, thus flattening structures in the redshift direction. The effect of this upon the correlation function is actually quite complicated and depends upon the direction cosine μ between the line of sight $\mathbf{l}$ and the separation $\mathbf{s}$. One can show, however, that the angle-averaged redshift space correlation function is given by the simple form

$$\bar{\xi}(s) = \left(1 + \frac{2f}{3} + \frac{f^2}{5}\right)\xi_r(s), \tag{6.48}$$

where ξ_r is the real space correlation function (Kaiser 1987). In principle, this equation allows one to estimate Ω (through the f dependence), but as usual one ends up with a determination of the parameter β defined in equation (6.32).

Perhaps a better way to use redshift-space distortions in the linear regime is to study their effect on the power-spectrum, where the directional dependence is easier to calculate. In fact, one can show quite easily that

$$P_s(\mathbf{k}) = P_r(\mathbf{k})[1 + (\mu f)^2], \tag{6.49}$$

where P_s and P_r are the redshift space and real space power-spectra respectively. If one can estimate the power-spectrum in various directions of $\mathbf{k}$, then one can fit the expected μ dependence to obtain an estimate of f and hence β.

Present estimates arising from this type of analysis are limited by the size of available galaxy catalogues and are subject to considerable uncertainties, but it does appear that they yield rather lower values of β than the peculiar-motion studies mentioned above: $\beta \sim 0.5$, rather than $\beta \sim 1$ (e.g. Fisher *et al.* 1994; Cole, Fisher & Weinberg 1995); the quoted values are for IRAS galaxies. Uncertainties are therefore very large (see Strauss

& Willick 1995), so one would not wish to bet one's house on these estimates being correct. Prospects for the redshift-space distortion analyses in the future will brighten when data from, in particular, Sloan and 2DF become available. Such studies should also remove doubts concerning how representative are the more local determinations.

6.5 Summary

A detailed list of determinations of β and Ω using the approaches we have discussed, along with others, is presented in Table 3 of Strauss & Willick (1995), from which one can gauge the scatter in the various determinations. We now summarise the main results and indicate which we believe to be the most compelling.

• Large-scale clustering statistics are open to different interpretations and do not provide direct constraints on Ω_0. As far as we are aware the data on power-spectra, abundances and so on are all consistent with the COBE normalisation for low-density CDM models (Liddle *et al.* 1996), although other possibilities of high-density universes with different matter content are not excluded.
• The behaviour of the dipole measured for IRAS galaxies seems to indicate a high value of β and thence, if one assumes that IRAS galaxies are relatively unbiased, a value of $\Omega_0 \simeq 1$. It is unclear, however, whether the available samples are sufficiently deep to ensure the convergence of the dipole, so this result is not watertight.
• It is fair to say that both the bulk flow data and the POTENT analysis are generally consistent with a high value of Ω_0. To accept these estimates, however, one needs to believe that the distance indicators are accurate to within, say, 10–20 %. It is yet unproven that the data are this reliable.
• Redshift-space anisotropies are in principle much more robust, as they do not depend on distance estimates, but they are statistical and therefore limited by the data available. There is some indication, however, that these studies produce $\beta \simeq 0.5$, rather than $\beta \simeq 1$ as in the peculiar motion studies. These will probably be the most reliable methods for determining Ω_0 from large-scale structure when the new generation of redshift surveys come on

line, as they are much more robust than methods based on distance estimates.

- The weak dependence of the growth factor f on Λ means that there is little hope of using large-scale dynamics to constrain Ω_0.
- *All* of these analyses are rendered questionable by the arbitrariness of the linear bias factor generally employed.

Consequent on the last comment, the most important future development in this area would be a believable theoretical understanding of the bias factor b and its possible dependence on cosmological parameters (including Ω_0). Until that is achieved, the results obtained from these methods, despite their intrinsic value and interest, will necessarily remain circumstantial rather than conclusive.

7
The cosmic microwave background

The detection of fluctuations in the sky temperature of the cosmic microwave background (CMB) by the COBE team (Smoot *et al.* 1992) was an important milestone in the development of cosmology. Aside from the discovery of the CMB itself, it was probably the most important event in this field since Hubble's discovery of the expansion of the universe in the 1920s (Hubble 1929). The importance of the COBE detection lies in the way these fluctuations are supposed to have been generated, and their relation to the present matter distribution. As we shall explain shortly, the variations in temperature are thought to be associated with density perturbations existing at the epoch $t_{\rm rec}$, when matter and radiation decoupled. If this is the correct interpretation, then we can actually look back directly at the power spectrum of density fluctuations at early times, before it was modified by non-linear evolution and without having to worry about the possible bias of galaxy power spectra.

The search for anisotropies in the CMB has been going on for around 25 years. As the experiments got better and better, and the upper limits placed on the possible anisotropy got lower and lower, theorists concentrated upon constructing models which predicted the smallest possible temperature fluctuations. The baryon-only models of the 1970s were discarded primarily because they could not be modified to produce low enough CMB fluctuations. The introduction of dark matter allowed such a reduction and the culmination of this process was the introduction of bias, which reduces the expected temperature fluctuation still further. It was an interesting experience to those who had been working in this field for many years to see this trend change sign abruptly in 1992. The $\Delta T/T$ fluctuations seen by COBE turned out to be larger than predicted by the standard version of the CDM model. This must have been the first time a theory had been rejected because it did not produce high enough temperature fluctuations!

Searches for CMB anisotropy are, on their own, enough subject matter for a whole book (e.g. Partridge 1995). In just one chapter we must therefore limit our scope quite considerably. Moreover, COBE marked the start, rather than the finish, of this aspect of cosmology and it is pointless to produce a definitive review of all the ongoing experiments and implications of the various upper limits and half-detections for specific theories, when it is possible (and indeed probable) that the whole picture will change within a year or two. We shall therefore mainly concentrate on trying to explain the physics responsible for various forms of temperature anisotropy, and the prospects for constraining the cosmological density parameter using appropriate observations. We shall not discuss any specific models in detail, except as illustrative examples, and our treatment of the experimental side of this subject will be brief and non-technical. Finally, we shall be extremely conservative when it comes to drawing conclusions. As we shall explain, the experimental situation with respect to CMB anisotropy as a function of angular scale is still very confused and we feel the wisest course is to wait until observations are firmly established before drawing definite conclusions.

7.1 Introduction

7.1.1 The relic radiation

As well as matter, the universe is filled with a thermal radiation background, called the *cosmic microwave background* (CMB) radiation. This was discovered by Penzias & Wilson (1965), for which they won the Nobel Prize. In fact this discovery was entirely serendipitous. Penzias and Wilson were radio engineers investigating the properties of atmospheric noise in connection with the Telstar communication satellite project. They found an apparently uniform background 'hiss' at microwave frequencies which could not be explained by instrumental noise or by any known radio sources. After careful investigations they admitted the possible explanation that they had discovered a thermal radiation background such as that expected to be left as a relic of the primordial fireball phase. In fact, the existence of a radiation background of roughly the same properties as that observed was predicted by George Gamow in the mid-1940s, but this prediction was not known to Penzias and Wilson. A group of theorists at Princeton University (Dicke *et al.* 1965) soon saw the possible interpretation of the background 'hiss' as relic radiation, and their paper was published alongside the Penzias & Wilson paper in the *Astrophysical Journal.*

The CMB radiation possesses a nearly perfect *black-body spectrum.* At the time of its discovery the CMB was known to have an approximately thermal spectrum, but other explanations were possible. Advocates of the steady state proposed that one was merely observing starlight reprocessed by dust, and models were constructed which accounted for the observations reasonably well. In the past 30 years, however, continually more sophisticated experimental techniques have been directed at the measurement of the CMB spectrum, exploiting ground-based antennae, rockets, balloons and, most recently and effectively, the COBE satellite. The COBE satellite had an enormous advantage over previous experiments: it was able to avoid atmospheric absorption, which plays havoc with ground-based experiments at microwave and submillimetre frequencies. The spectrum supplied by COBE reveals just how close to an ideal black body† the radiation background is; the temperature of the CMB is now known to be 2.726 ± 0.005 K. We discussed in Chapter 5 how the lack of any apparent distortion of the observed spectrum from the ideal black-body shape places constraints on the contribution to Ω_0 from hot gas. In the context of this chapter, however, we should note that the cosmic microwave background contributes directly to the energy density of the universe an amount corresponding to

$$\rho_{0r} = \frac{\sigma T_{0r}^4}{c^2} \simeq 4.8 \times 10^{-34} \text{ g cm}^{-3}, \tag{7.1}$$

where $\sigma = \pi^2 k_B^4 / 15\hbar^3 c^3$ is the black-body constant; the Stefan–Boltzmann constant is just $\sigma c/4$). The corresponding contribution

† Attempts to account for this in a steady-state model by non-thermal processes seem entirely contrived. The CMB radiation really is good evidence that the big-bang model is correct.

to the density parameter is

$$\Omega_r \simeq 2.3 \times 10^{-5} h^{-2}. \tag{7.2}$$

This is much smaller than the matter contribution, even if we include only the baryonic part discussed in Chapter 4. The smallness of the number (7.2) masks the fact that the CMB radiation is potentially an extremely important diagnostic for Ω_0, not because of its direct contribution to the mean energy density but because of the expected properties of the small fluctuations in temperature with position on the sky.

7.1.2 The angular power spectrum

For the rest of this chapter, we shall concentrate on the temperature anisotropy of the CMB, and its implications for cosmological models. Let us first describe the statistical characterisation of fluctuations in the temperature of the CMB radiation from point to point on the celestial sphere.

The usual approach, which we shall follow here, is to expand the distribution of T on the sky as a sum over spherical harmonics

$$\frac{\Delta T(\theta,\phi)}{T} = \sum_{l=0}^{\infty} \sum_{m=-l}^{m=+l} a_{lm} Y_{lm}(\theta,\phi), \tag{7.3}$$

where θ and ϕ are the usual spherical angles; $\Delta T/T$ is defined by

$$\frac{\Delta T(\theta,\phi)}{T} = \frac{T(\theta,\phi) - \bar{T}}{\bar{T}} \tag{7.4}$$

and $\bar{T}$ is the average temperature on the sky. The $l = 0$ term is a monopole correction which essentially just alters the mean temperature on a particular observer's sky with respect to the global mean over an ensemble of all possible such skies. We shall ignore this term from now on because it is not measurable. The $l = 1$ term is a dipole term which, as we discussed in §6.3, is attributable† to our motion through space. Since this anisotropy is presumably generated locally by matter fluctuations, one tends to remove the $l = 1$ mode and treat it separately. The remaining modes, from the quadrupole ($l = 2$) upwards, are usually attributed to intrinsic

† Although not uniquely: the dipole could also, in principle, be generated by inhomogeneity. The usual interpretation can in principle be tested by looking for anisotropic number counts; see Ellis & Baldwin (1984).

anisotropy produced by effects either at $t_{\rm rec}$ or between $t_{\rm rec}$ and t_0. For these effects the sum in equation (7.3) is generally taken over $l \geq 2$. Higher l modes correspond to fluctuations on smaller angular scales ϑ according to the approximate relation

$$\vartheta \simeq \frac{60^\circ}{l} \,. \tag{7.5}$$

The expansion of $\Delta T/T$ in spherical harmonics is entirely analogous to the plane-wave Fourier expansion of the density perturbations δ; cf. equation (6.14); the Y_{lm} are a complete orthonormal set of functions on the surface of a sphere, just as the plane wave modes are a complete orthonomal set in a flat three-dimensional space. The a_{lm} are generally complex, and satisfy the conditions

$$\langle a^*_{l'm'} a_{lm} \rangle = C_l \delta_{ll'} \delta_{mm'}, \tag{7.6}$$

where δ_{ij} is the Kronecker symbol and the average is taken over an ensemble of realisations. The quantity C_l is the *angular power spectrum*,

$$C_l \equiv \langle |a_{lm}|^2 \rangle, \tag{7.7}$$

which is analogous to the power spectrum $P(k)$ defined by equation (6.15). It is also useful to define an *autocovariance function* for the temperature fluctuations,

$$C(\vartheta) = \left\langle \frac{\Delta T}{T}(\hat{\mathbf{n}}_1) \frac{\Delta T}{T}(\hat{\mathbf{n}}_2) \right\rangle, \tag{7.8}$$

where

$$\cos\vartheta = \hat{\mathbf{n}}_1 \cdot \hat{\mathbf{n}}_2 \tag{7.9}$$

and the $\hat{\mathbf{n}}_i$ are unit vectors pointing to arbitrary directions on the sky. The expectation values in (7.7) and (7.8) are notionally taken over an ensemble of all possible skies. In practice, we only have one sky available for observation, so C_l and $C(\vartheta)$ are not, in themselves, observable quantities. One can try to estimate C_l or $C(\vartheta)$ from an individual sky using an *ergodic hypothesis*: an average over the probability ensemble is the same as an average over all spatial positions within a given realisation. This only works on small angular scales when it is possible to average over many different pairs of directions with the same ϑ, or many different modes with the same l. On larger scales, however, it is extremely difficult to estimate the true $C(\vartheta)$ because there are so few independent directions at large ϑ or, equivalently, so few independent l modes at small l. Large-angle statistics are therefore complicated by the fact that we inhabit one realisation and there is no reason why this should possess exactly the ensemble average values of the relevant quantities.

As was the case with the spatial power spectrum and covariance functions, there is a simple relationship between the angular power spectrum and covariance function:

$$C(\vartheta) = \frac{1}{4\pi}\sum_{l=2}^{\infty}(2l+1)C_l P_l(\cos\vartheta), \tag{7.10}$$

where $P_l(x)$ is a Legendre polynomial. We have written the sum explicitly to emphasise our omission of the monopole and dipole contributions from (7.10) and the following.

It is quite straightforward to calculate the cosmic variance corresponding to an estimate obtained from observations of a single sky, $\hat{C}(\vartheta)$, of the 'true' autocovariance function, $C(\vartheta)$:

$$\hat{C}(\vartheta) = \frac{1}{4\pi}\sum_{l=2}^{\infty}\sum_{m=-l}^{l}|\hat{a}_{lm}|^2 P_l(\cos\vartheta), \tag{7.11}$$

where $\hat{a}_{lm}$ are obtained from a single realisation on the sky. The statistical procedure for estimating these quantities is by no means trivial, but we shall not describe the various possible approaches here. In fact, the variance of the estimated $\hat{a}_{lm}$ across an ensemble of skies will be $|a_{lm}|^2$ so that the $\hat{C}(\theta)$ will have variance

$$\langle|\hat{C}(\vartheta) - C(\vartheta)|^2\rangle = \left(\frac{1}{4\pi}\right)^2\sum_{l=2}^{\infty}(2l+1)C_l^2 P_l^2(\cos\vartheta). \tag{7.12}$$

In the following sections we shall discuss the various physical processes that produce anisotropy with a given form of C_l, excluding the dipole. Generally the form of C_l must be computed numerically, at least on small and intermediate scales, by solving the transport equations for the matter-radiation fluid through decoupling. We shall make some remarks on how this is done later in this chapter. As we shall see, the comparison of a theoretical C_l against an observed $\hat{C}_l$ or $\hat{C}(\vartheta)$ in principle provides a powerful test of theories of galaxy formation. The left-hand panel of Figure 7.1 shows the behaviour of C_l expected in some illustrative models of galaxy formation which we shall discuss later on in this chapter; the right-hand figure shows various experimental

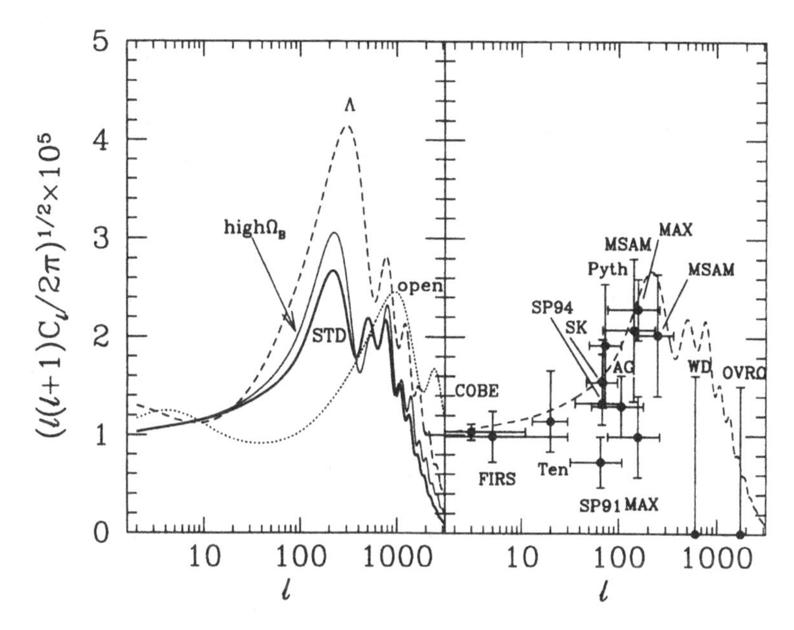

Fig. 7.1 Theoretical angular spectra of the CMB radiation (left), along with experimental detections and upper limits (right) which we discuss later. On the left, STD refers to the 'standard' CDM model with $\Omega_0 = 1$, $h = 0.5$ and $\Omega_b = 0.05$. CDM models with $\Omega_0 = 0.1$ both with (dashed) and without (dotted) the Λ term are also shown. A model with $\Omega_b = 0.1$ is also shown, slightly above the STD curve. Picture courtesy of Naoshi Sugiyama.

detections and/or upper limits plotted at the l-values where the appropriate F_l are most sensitive and with horizontal error-bars roughly indicating the range of l that the experiments cover.

7.1.3 *Folklore and phenomenology*

In 'standard' models of structure formation, described in the last chapter, small primordial perturbations to the energy density are amplified by gravitational instability as the universe expands. Although these fluctuations were small at the time of recombination, they are predicted to leave a detectable imprint in the CMB radiation (Sachs & Wolfe 1967; Hogan, Kaiser & Rees 1982; Kaiser & Silk 1987). The COBE experiment has indeed detected fluctuations in the temperature of the CMB on the sky of roughly the right amplitude to be consistent with this standard picture. The issue is, what do the COBE results mean for the value of Ω_0?

Perhaps the strongest claims of observational evidence against a low density universe are made in connection with the measured microwave background radiation anisotropy. The claim is that in low density universes the low degree of this anisotropy contradicts the requirements of structure formation – basically because growth of inhomogeneities freezes out at an early time in low density universes. So in such universes larger initial inhomogeneities are required which, it is claimed, lead to a higher CMB anistropy than is observed. The argument is essentially that, as we explained in Chapter 6, in an Einstein–de Sitter universe, linear gravitational instability proceeds in such a way that the density contrast δ grows linearly with the cosmic scale factor $a(t)$. In other words, the density contrast scales with redshift as $(1 + z)^{-1}$. There is also a simple relationship between a given (comoving) length scale and the angle it subtends on the celestial sphere at recombination (see below). This, in principle, allows a relatively straightforward comparison to be made between the temperature fluctuations on a given angular scale to the density fluctuations on a given length scale.

At early times, the linear growth of fluctuations in an open universe is indistinguishable from the Einstein–de Sitter case. At later times, however, the expansion becomes curvature-dominated and, under these circumstances, density fluctuations can no longer grow at all. In order to get a given level of fluctuation at the present time ($z = 0$) one would seem to require that density fluctuations at recombination need be a larger than in the flat case by a factor which depends quite strongly on Ω_0. Since the level of fluctuations observed is of the order of that expected in flat models of the universe, it would appear that models with Ω_0 much less than unity are excluded by CMB observations.

7.1.4 Compatible low-density models

At a certain level the preceding argument is disproved by the fact that several well-established and detailed technical studies (Wilson 1983; Gouda, Sugiyama & Sasaki 1991a,b; Wright *et al.*

1992; Kamionkowski & Spergel 1994; Ratra *et al.* 1996; Cayón *et al.* 1996) have provided low-density models that are broadly consistent with the COBE measurements (Smoot *et al.* 1992; Bennett *et al.* 1996). However, the belief that the microwave background anistropy excludes low-density universes appears to be widespread, so it is worth examining in detail why this is not the case.

7.2 The Sachs–Wolfe effect

We begin by describing the Sachs–Wolfe effect (Sachs & Wolfe 1967), because the physical origin of this effect is relatively easy to understand and, more importantly, the fluctuations detected by COBE are usually interpreted as being due to it: the Sachs–Wolfe effect is expected to be the dominant contribution to $\Delta T/T$ on large angular scales. Basically, this is a relativistic effect due to the fact that photons travelling to an observer from the last scattering surface encounter metric perturbations which cause them to change frequency. One can understand this in a Newtonian context by noting that metric perturbations correspond to perturbations in the gravitational potential, $\delta\varphi$, in Newtonian theory, and these, in turn, are generated by density fluctuations, $\delta\rho$. Photons climbing out of such potential wells suffer a gravitational redshift but also a time dilation effect so that one effectively sees them at a different time, and thus at a different value of a, to unperturbed photons. In a flat, pressureless universe with $\Lambda = 0$, the first effect gives

$$\frac{\Delta T}{T} = \frac{\delta\varphi}{c^2}\,, \tag{7.13}$$

while the second contributes

$$\frac{\Delta T}{T} = -\frac{\delta a}{a} = -\frac{2}{3}\frac{\delta t}{t} = -\frac{2}{3}\frac{\delta\varphi}{c^2}\;: \tag{7.14}$$

the net effect is therefore

$$\frac{\Delta T}{T} = \frac{1}{3}\frac{\delta\varphi}{c^2} \simeq \frac{1}{3}\frac{\delta\rho}{\rho}\left(\frac{\lambda}{ct}\right)^2, \tag{7.15}$$

where λ is the scale of the perturbation.

This is the case for initially adiabatic fluctuations. Since the Sachs–Wolfe effect is generated by fluctuations in the metric, one might expect that the alternative isocurvature fluctuations (perturbations in the entropy which leave the energy density unchanged and therefore correspond to negligible fluctuations in the metric) should produce a very small Sachs–Wolfe anisotropy. This is not the case, for two reasons. Firstly, isocurvature fluctuations do generate significant fluctuations in the matter component and hence in the gravitational potential, when they enter the horizon; this is due to the influence of pressure gradients. In addition, isocurvature fluctuations generate significant fluctuations in the radiation density after $z_{\rm eq}$, the redshift corresponding to matter–radiation equality, because the initial entropy perturbation is then transferred into the perturbation of the radiation. The total anisotropy seen is therefore the sum of the Sachs–Wolfe contribution and the intrinsic anisotropy carried by the radiation. The upshot of all this is that the net anisotropy seen is a factor of six larger for isocurvature fluctuations than for adiabatic ones. This is sufficient on its own to rule out most isocurvature models since the level of anisotropy detected is roughly that expected for adiabatic perturbations (Efstathiou & Bond 1986, 1987).

According to equation (7.15), the temperature anisotropy is produced by gravitational potential fluctuations sitting on the last scattering surface. In fact this is not quite correct, and there are actually two other contributions arising from the Sachs–Wolfe effect. We shall discuss the second of these in §7.2.2 below. The first is a term

$$\frac{\Delta T}{T} \simeq 2 \int \frac{\delta\varphi^{\cdot}}{c^2}\, \mathrm{d}t, \tag{7.16}$$

where the integral is taken along the path of a photon from the last scattering surface to the observer. We discuss the importance of this effect on large angular scales for open models in particular, in §7.2.3; it can also be relevant on smaller angular scales. Many writers distinguish this effect from the Sachs–Wolfe effect by calling it the *integrated* Sachs–Wolfe effect or even the *Rees–Sciama effect* (Rees & Sciama 1968). We disagree with this because the contribution (7.16) is actually included in the generic Sachs–Wolfe formula and there seems to us to be no point in adding yet another name to the lexicon. The physical origin of this effect is the change in depth of a potential well as a photon crosses it. If the well does not deepen, a photon does not suffer a net shift in energy from falling in and then climbing up. If the potential changes while the photon moves through it, however, there will be a net change in the frequency. In a flat universe, $\delta\varphi$ is actually constant in linear theory, so one needs to have non-linear evolution in order to produce a non-zero Sachs–Wolfe effect. Since the potential fluctuations are of the order of $\delta\varphi \simeq \delta(\lambda/ct)^2$ one requires significant non-linear evolution of δ on very large scales to obtain a reasonably large contribution. To calculate the effect in detail for a background of perturbations is quite difficult because of the inherent non-linearity involved, though it has been attempted in inhomogeneous Swiss-cheese models (e.g. Dyer 1976). It turns out that, for a spherical void of the same diameter as a large void seen in Boötes, one expects to see a cold spot corresponding to $\Delta T/T \simeq 10^{-7}$ on an angular scale around 15°. The Shapley concentration of clusters is expected to produce a hotspot with $\Delta T/T \simeq 10^{-5}$ on a scale around 20°. In general these effects are smaller than the intrinsic CMB anisotropies we described above but may be detectable in large, sensitive sky maps: the position on the sky of these features should correspond to known features of the galaxy distribution.

7.2.1 The flat case

For the moment, we shall assume that we are dealing with temperature fluctuations produced by the potential wells produced by density fluctuations with a power spectrum of the form $P(k) \propto k^n$; cf. equation (6.19). What is the form of C_l predicted for fluctuations generated by this effect? This can be calculated quite straightforwardly by writing $\delta\varphi$ as a Fourier expansion and using the fact that the power spectrum of $\delta\varphi$ is proportional to $k^{-4}P(k)$, where $P(k)$ is the power spectrum of the density fluctuations. Expanding the net $\Delta T/T$ in spherical harmonics and averaging over all possible observer positions yields, after some work,

$$C_l = \langle |a_{lm}|^2 \rangle = \frac{1}{2\pi}\left(\frac{H_0}{c}\right)^4 \int_0^\infty P(k) j_l^2(kx) \frac{\mathrm{d}k}{k^2}, \tag{7.17}$$

where j_l is a spherical Bessel function and $x = 2c/H_0$. For an initial power spectrum of the form $P(k) \propto k$, the quantity $l(l+1)C_l$ is independent of the mode order l for Sachs–Wolfe perturbations. This is shown in the left panel of Figure 7.1: the shape of C_l for small l is determined purely by the shape of $P(k)$, the shape of the primordial fluctuation spectrum before it is modified by the transfer function. The reason for this is easy to see: the scale of the horizon at $z_{\rm rec}$ is of the order of

$$\vartheta_H(z_{\rm rec}) \simeq \left(\frac{\Omega}{z_{\rm rec}}\right)^{1/2} \text{ radians,} \tag{7.18}$$

so that $\vartheta_H \simeq 2°$ for $z_{\rm rec} \simeq 1000$, which is the usual situation. Fluctuations on angular scales larger than this will retain their primordial character since they will not have been modified by any causal processes inside the horizon before $z_{\rm rec}$. One must therefore be seeing the primordial (unprocessed) spectrum. This is particularly important because observations of C_l at small l can then be used to normalise $P(k)$ in a manner independent of the shape of the power spectrum, and therefore independent of the nature of the dark matter.

One simple way to do this is to use the quadrupole perturbation modes which have $l = 2$. There are five spherical harmonics with $l = 2$, so the quadrupole has five components a_{2m} ($m = -2, -1, 0, 1, 2$) that can be determined from a map of the sky even if it is noisy. From (7.17), we can show that, if $P(k) \propto k$, then

$$\langle |a_{2m}|^2 \rangle = C_2 \simeq \frac{\pi}{3}\left(\frac{H_0 R}{c}\right)^4 \left(\frac{\delta M}{M}\right)_R^2. \tag{7.19}$$

This connects the observed temperature pattern on the sky with the mass fluctuations $\delta M/M = \sigma_M$ observed at the present epoch on a scale R; cf. equation (6.18). Note that this connection is independent of the *source* of the initial density fluctuations. In particular it does not require them to have been generated during an inflationary epoch.

7.2.2 *Gravitational waves*

The second additional contribution comes from tensor metric perturbations, i.e. *gravitational waves.* These do not correspond to density fluctuations and have no Newtonian analogue but they do produce redshifting as a result of the perturbations in the metric. As we shall see at the end of this section, gravitational waves capable of generating large-scale anisotropy of this kind are predicted in many inflationary models so this is potentially an important effect.

What is the possible contribution of tensor perturbation modes to the large-scale CMB anisotropy? Gravitational waves do involve metric fluctuations and therefore do generate a Sachs–Wolfe effect on scales larger than the horizon. Once inside the horizon, however, they redshift away (just like relativistic particles) and play no role at all in structure formation. Clearly then, normalising the power spectrum $P(k)$ to the observed C_l using (7.17) is incorrect if the tensor signal is significant (e.g. Lidsey & Coles 1992).

One can define a power spectrum of gravitational wave perturbations in an analogous fashion to that of the density perturbations. It turns out that inflationary models also generically predict a tensor spectrum of power-law form, but with a spectral index

$$n_T = 1 - 2\epsilon_*, \tag{7.20}$$

instead of the $n_T = 1$ expected on naive grounds (e.g. Liddle & Lyth 1993). Since ϵ_* is a small parameter, the tensor spectrum will be close to scale invariant. It is also possible to calculate the ratio, $\mathcal{R}$, between the tensor and scalar contributions to C_l:

$$\mathcal{R} = \frac{C_l^{\mathrm{T}}}{C_l^{\mathrm{S}}} \simeq 12\epsilon_*. \tag{7.21}$$

To get a significant value of the gravitational wave contribution to C_l one therefore generally requires a significant value of ϵ_* and therefore both scalar and tensor spectra will usually be expected to be tilted away from $n = 1$. A scalar spectrum with $n = 0.85$ would make the ratio (7.21) equal to unity. If $\mathcal{R} = 1$, then one can reconcile the COBE detection with a CDM model having a significantly high value of b. Because one cannot use Sachs–Wolfe anisotropies alone to determine the value of $\mathcal{R}$, there clearly remains some element of ambiguity in the normalisation of $P(k)$ to the COBE results.

Equations (7.20) and (7.21) are true for inflationary models with a single scalar field. More contrived models with several scalar fields can allow the two spectral indices and the ratio to be given essentially independently of each other. The shape of the COBE autocovariance function suggests that n cannot be much less than unity, so the prospects for having a single-field inflationary model producing a large tensor contribution seem small. On the other hand, we have no *a priori* information about the value of $\mathcal{R}$, so it would be nice to be able to constrain it using observations. It turns out that to perform such a test requires, at the very least, observations on a different (i.e. smaller) angular scale: while the scalar contribution increases around degree scales, the tensor contribution dies away completely. In principle, one can therefore estimate $\mathcal{R}$ by comparing observations of C_l at different values of l although, as we shall see, the result depends on other cosmological parameters. Note, however, that in contrast to relation (7.19), the formulae discussed here are dependent on an a specific inflationary source for the perturbations.

7.2.3 Non-flat models

We now return to the argument given in §7.1.3, which suggested that one should see much larger radiation anisotropies in an open universe. As we explained in Chapter 6, fluctuation growth effectively freezes out in open low-density universes at a redshift z given by $1 + z \simeq 1/\Omega_0$ because at late times the expansion becomes dominated by the curvature term in the Friedman equation. More accurately, the growth of fluctuations in an open universe is a factor $g(\Omega_0) \simeq \Omega_0^{0.65}$ smaller in an open universe than in a flat universe (e.g. Peacock & Dodds 1994). Together with the fact that the density itself is a factor Ω_0 less than in the flat case, this implies that there should be a factor g^2/Ω_0^2 at the front of the right-hand side of equation (7.17). One might therefore expect the CMB fluctuations on a given scale to be a factor $\Omega_0/g \simeq \Omega_0^{-0.35}$ larger than in a flat universe. This is not, in fact, the case for several reasons.

First, in a low density universe the relation between observational angles and corresponding linear scales in the last scattering surface is also altered significantly from that holding in a critical density universe. Roughly speaking, the angle θ subtended by a comoving length scale l_0 Mpc is given by

$$\theta(l_0) \simeq \frac{\Omega_0}{2c} l_0 H_0, \tag{7.22}$$

so an observation at a given angular scale provides a constraint on the fluctuations on a larger length scale in the open case. More accurate results are given by Ellis & Tavakol (1993) and

Kamionkowski, Spergel & Sugiyama (1994). The net effect of lowering Ω_0 depends on the relative amplitudes of fluctuations on different length scales, i.e. on the power spectrum of the initial perturbations. For the usual Harrison–Zel'dovich scale-free spectrum of initial fluctuations, there is no dependence of the temperature fluctuations induced by the Sachs–Wolfe effect on the length scale of the perturbation responsible. In this case the factor $\Omega_0^{0.35}$ of what would have been expected in a critical density universe does indeed apply. In other circumstances, the result can be a lower CMB anisotropy than in the critical density cases, on larger angular scales; see Traschen & Eardley (1986).

The presence of spatial curvature in open universes also poses problems for the interpretation of these results: the notion of scale invariance becomes complicated in open models, because there is a characteristic scale: the curvature scale. Moreover, in negatively curved spaces, the plane wave Fourier expansion used to derive equation (7.17) is not appropriate (Wilson 1983; Gouda, Sugiyama & Sasaki 1991a,b). One cannot simply therefore translate the flat space results for a given power spectrum to the case of a curved space without modification.

Another complicating factor involves the role of pressure. The usual relation between density inhomogeneities at last scattering and the microwave background radiation is calculated in a critical density universe, where matter dominates the dynamics at the time of decoupling. In a low density universe, however, pressure effects may be more significant at that time and, in principle, might alter the relation. The magnitude of this effect can be calculated directly by integrating the conservation equations along the null cone (Stoeger, Ellis & Xu 1994). The conclusion emerging from this approach is that there can be a considerable suppression of the amplitude of the temperature anisotropies in low density universes, compared with what one would expect simply on the basis of the standard Sachs–Wolfe formula. It must be emphasised that this result is robust in that (a) it does not depend on the usual splitting into scalar, vector and tensor modes, which is problematic because of its non-local nature, (b) it does not depend on the neglect of decaying perturbation modes, which are potentially significant in evaluating the background radiation, and (c) in this relation, the matter is treated as baryonic (i.e.

cold); any hot component will result in even further suppression. Moreover, in models with spatial curvature and/or a cosmological constant there are contributions to the temperature anistropy from the time-dependent Sachs–Wolfe effect which are absent in the flat matter-dominated case, and which we discuss in the next section. It appears that this effect is relatively well described by an additional factor of $g^{0.7}$ in the inferred power spectrum, so that the implied amplitude of density fluctuations scales only as $\Omega_0^{-0.12}$ (White & Bunn 1995). On the other hand, if there is a cosmological constant with flat spatial sections then the dependence is as $\Omega_0^{-0.69}$: much higher temperature fluctations are induced in this case.

These complications demonstrate that one really has to perform calculations very carefully, even on large angular scales where the physics appears to be relatively simple. Calculations which have been formed including all the relevant factors we have discussed confirm the claim we made in §7.1.3: that there is no problem in practice in finding models that match the growth of structure in open universes with the amplitude of the power spectrum inferred from the COBE results.

7.3 Smaller-scale radiation anisotropies

As we have already explained, the large-scale features of the microwave sky are expected to be primordial in origin. Smaller scales are closer to the size of the Hubble horizon at $z_{\rm rec}$, so the density fluctuations present there may have been modified by various damping and dissipation processes. Moreover, there are physical mechanisms other than the Sachs–Wolfe effect which are capable of generating anisotropy in the CMB on these smaller scales.

Let us begin with some naive estimates. For a start, if the density perturbations are adiabatic, then one should expect fluctuations in the photon temperature of the same order. Using $\rho_r \propto T^4$ and the adiabatic condition, $4\delta_m = 3\delta_r$, we find that

$$\frac{\Delta T}{T} \simeq \frac{1}{3}\frac{\delta\rho}{\rho}, \tag{7.23}$$

which is also stated implicitly above. Another mechanism, first discussed by Sunyaev & Zel'dovich (1970), is simply a Doppler effect. Density perturbations at $t_{\rm rec}$ will, by the continuity equation, induce streaming motions in the plasma. This generates a temperature anisotropy because some electrons are moving towards the observer when they last scatter the radiation and some are moving away. It turns out that the magnitude of this effect for perturbations on a scale λ at time t is

$$\frac{\Delta T}{T} \simeq \frac{\delta \rho}{\rho}\left(\frac{\lambda}{ct}\right), \tag{7.24}$$

where ct is of the order of the horizon scale at t.

The actual behaviour of the background radiation spectrum is, however, much more complicated than these simple arguments might suggest. The detailed computation of fluctuations originating on these scales is consequently much less straightforward than was the case for the Sachs–Wolfe effect. In general one therefore resorts to a full numerical solution of the Boltzmann equation for the photons through recombination, taking into account the effect of Thomson scattering. Some progress can be made, however, by looking at a simplified model: that of two interacting fluids.

7.3.1 A two-fluid model

After recombination, and the consequent decoupling of matter and radiation, the perturbations $\delta\rho_m$ in the total matter density evolve in the same way regardless of whether they were originally of adiabatic or isothermal type. Because there is essentially no interaction between the matter and radiation, and the radiation component is dynamically negligible compared to the matter component, the universe behaves as a single-fluid dust model. For the period prior to recombination and decoupling, it is more accurate to consider the matter and radiation components as two fluids interacting with each other on characteristic time-scales $\tau_{e\gamma}$ and $\tau_{\gamma e}$.

A detailed analysis of this two-fluid model (Coles & Lucchin 1995, §12.9) shows that, in general, before decoupling there are two fluctuation modes of approximately adiabatic nature, in the sense that $\delta_r/\delta_m \simeq 4/3$. These modes are unstable for wavelengths λ greater than the adiabatic Jeans length λ_J so that one mode increases with time and the other decreases; for $\lambda < \lambda_J$ they evolve like damped acoustic waves with the adiabatic sound speed v_s. A third mode, again of approximately adiabatic type, also exists but is non-propagating and always damped. The fourth and final mode is approximately isothermal (in the sense that $|\delta_r| \ll |\delta_m|$), so that for $\lambda > \lambda_J$ it is an unstable growing mode, but with a characteristic growth time $\tau > \tau_H$, so it is effectively frozen in. During decoupling, the last two of these modes gradually transform themselves into two isothermal modes which oscillate like waves for $\lambda < \lambda_J$, and are unstable (one growing and the other decaying) for $\lambda > \lambda_J$. The first two modes become purely radiative, i.e. $\delta_m \simeq 0$, and are unstable for wavelengths greater than the appropriate Jeans length for radiation $\lambda_J^{(r)}$; they oscillate like waves propagating at a speed $c/\sqrt{3}$ practically without damping for $\lambda < \lambda_J^{(r)}$. These last two modes are actually spurious, since in reality the radiation after decoupling behaves like a collisionless fluid which cannot be described by the hydrodynamical equations. A more exact treatment of the radiation shows that, for $\lambda > \lambda_J^{(r)}$ and after decoupling, there is a rapid damping of these purely radiative perturbations owing to the free streaming of photons whose mean free path is $l_\gamma \gg \lambda$.

One important outcome of this treatment is that, in general, the four modes correspond neither to purely adiabatic nor to purely isothermal modes. A generic perturbation must be thought of as a combination of four perturbations, each one in the form of one of these four fundamental modes. Given that each mode evolves differently, the nature of the perturbation must change with time; one can, for example, begin with a perturbation of pure adiabatic type which, in the course of its evolution, assumes a character closer to a mode of isothermal type, and vice versa. One can attribute this phenomenon to the continuous exchange of energy between the various modes.

There are three main problems with this hydrodynamical approach. First, being essentially Newtonian, it does not take into account all necessary relativistic corrections. One cannot trust the results obtained with these equations on scales comparable to, or greater than, the scale of the cosmological horizon. Secondly, the description of the radiation as a fluid is satisfactory on length scales $\lambda \gg c\tau_{\gamma e}$ and for epochs during which $\tau_{\gamma e}(\tau_{e\gamma}) \ll \tau_H$: these conditions are true only for $z \gg z_{\rm rec}$. For later times, or for smaller scales, it is necessary to adopt an approach based on kinetic theory; we shall describe this kind of approach in the next section.

The last problem we should mention is that the approximations used to derive the dispersion relation (which yields the time dependence of the various modes) from the fluid equations are only acceptable for $z > z_{\rm rec}$. Clearly then, for a detailed comparison with observations we need a more accurate formalism.

7.3.2 Kinetic theory

The full numerical solution of the system of fully relativistic equations describing the matter and radiation perturbations (in a kinetic approach), and the perturbations in the spatial geometry (i.e. metric perturbations), is more complex still. Such computations enable one to calculate with great accuracy, given generic initial conditions at the entry of a baryonic mass scale in the cosmological horizon, the detailed behaviour of $\delta_m(M)$, as well as the perturbations to the radiation component and hence the associated fluctuations in the cosmic microwave background. As we have already mentioned, the exact relativistic treatment of the evolution of cosmological perturbations is very complicated. One must keep track not only of perturbations to both the matter and radiation but also of fluctuations in the metric. The Robertson–Walker metric describing the unperturbed background must be replaced by a metric whose components g'_{ik} differ by infinitesimal quantities from the original g_{ik}: the deviations δg_{ik} are connected with the perturbations to the matter and radiation by the Einstein equations. There is also the problem of choice of *gauge*. This is a subtle problem whose discussion we shall defer until the end of this section. The simplest approach† is to adopt the *synchronous gauge* characterised by the metric

$$ds^2 = (c\mathrm{d}t)^2 - a^2[\gamma_{\alpha\beta} - h_{\alpha\beta}(\mathbf{x}, t)]\mathrm{d}x^\alpha \mathrm{d}x^\beta, \qquad (7.25)$$

where $|h_{\alpha\beta}| \ll 1$. The treatment is considerably simplified if the unperturbed metric is flat so that $\gamma_{\alpha\beta} = \delta_{\alpha\beta}$, where $\delta_{\alpha\beta}$ is the Kronecker symbol: $\delta_{\alpha\beta} = 1$ for $\alpha = \beta$, $\delta_{\alpha\beta} = 0$ for $\alpha \neq \beta$. This is also the case in an approximate sense if the universe is not flat,

† This is the coordinate choice we made in Chapter 6. It was also the approach adopted in the classic paper of Peebles & Yu (1970), upon which much of this discussion is based.

but one is looking at scales much less than the curvature radius or at very early times, which is usually the case.

The time evolution of the trace h of the tensor $h_{\alpha\beta}$ is related to the evolution of matter and radiation perturbations through the Einstein equations in the form

$$\ddot{h} + 2\frac{\dot{a}}{a}\dot{h} = 8\pi G(\rho_m\delta_m + 2\rho_r\delta_r). \tag{7.26}$$

The equations that describe the evolution of the time-dependent parts δ_m and V_m of the perturbations in the density and velocity of the matter are

$$\begin{aligned} \dot{\delta}_m + ikV_m &= \frac{1}{2}\dot{h}, \\ \dot{V}_m + \frac{\dot{a}}{a}V_m + \frac{V_m - V_r}{\tau_{e\gamma}} &= 0; \end{aligned} \tag{7.27}$$

the perturbation in the velocity of the radiation V_r will be defined a little later.

As far as the radiation perturbations are concerned, one can demonstrate that their evolution is described by a single equation involving the *brightness function* $\delta^{(r)}(\mathbf{x}, t)$ or, more conveniently, its Fourier transform $\delta_k^{(r)}$. The quantity

$$\delta_r(k,t) = \frac{1}{4\pi}\int \delta_k^{(r)}(\vartheta, \varphi, t)d\Omega \tag{7.28}$$

at any point involves contributions from photons with momenta directions specified by the spherical polar angles ϑ and φ. The differential equation which describes the evolution of $\delta_k^{(r)}$, which was first derived from the Liouville equation by Peebles & Yu (1970), is

$$\dot{\delta}_k^{(r)} + ikc\cos\vartheta\,\delta_k^{(r)} + \frac{1}{\tau_{\gamma e}}\left(\delta_r + 4\frac{V_m}{c}\cos\vartheta - \delta_k^{(r)}\right) = 2\cos^2\vartheta\,\dot{h}, \tag{7.29}$$

where ϑ is the angle between the photon momentum and the wave vector $\mathbf{k}$, which we assume to define the polar axis of a local coordinate system. For a single plane wave, one can expand $\delta_k^{(r)}$ in angular moments σ_l defined with respect to the Legendre polynomials

$$\delta_k^{(r)} = \sum_l (2l+1)P_l(\cos\vartheta)\sigma_l(k,t). \tag{7.30}$$

The perturbation δ_r coincides with the moment σ_0, while the velocity perturbation V_r which appears in (7.27) is given by $\sigma_1/4$.

It is comparatively straightforward to show that the evolution of the brightness function is governed by a hierarchy of equations for the moments σ_l:

$$\dot{\sigma}_0 + ik\sigma_1 = \frac{2}{3}\dot{h}, \tag{7.31}$$

$$\dot{\sigma}_1 + ik\left(\frac{2}{3}\sigma_2 + \frac{1}{3}\sigma_0\right) = \frac{4}{3}\frac{V_m - V_r}{\tau_{\gamma e}}, \tag{7.32}$$

$$\dot{\sigma}_2 + ik\left(\frac{3}{5}\sigma_3 + \frac{2}{5}\sigma_1\right) = \frac{4}{15}\dot{h} - \frac{3\sigma_2}{4\tau_{\gamma e}}, \tag{7.33}$$

$$\dot{\sigma}_l + ik\left(\frac{l+1}{2l+1}\sigma_{l+1} + \frac{l}{2l+1}\sigma_{l-1}\right) = -\frac{\sigma_l}{\tau_{\gamma e}}, \tag{7.34}$$

for $l = 0$, $l = 1$, $l = 2$ and $l \geq 3$ respectively. One can verify that the two-fluid approximation practically coincides with the system of equations (7.31)–(7.34) if one puts $\sigma_3 = 0$ and neglects $\dot{\sigma}_2$ in equation (7.35), thus truncating the hierarchy. This approximation is good in the epoch during which $\tau_{\gamma e} \ll \tau_H$, which is in practice any time prior to recombination, and on large scales, such that $\lambda \gg c\tau_{\gamma e}$.

The appearance of a term in $\dot{h}$ in the $l = 2$ equation, which might appear surprising, originates from a term involving the local *shear* of the metric, but which has been eliminated using the $h_{0\beta}$ Einstein equation to relate it to the trace h.

In the general situation, both during and after recombination, the system can be solved only by truncating the hierarchy at some suitably high value of l; the number of l-modes one has to take grows steadily as decoupling and recombination proceed. In the opposite limit to that of the validity of the two-fluid approach, one has $\tau_{\gamma e} \gg \tau_H$, which is much later than recombination or for small scales such that $\lambda \ll c\tau_{\gamma e}$. In such a case we have

$$\dot{\delta}_k^{(r)} + ikc\cos\vartheta\,\delta_k^{(r)} = 2\cos^2\vartheta\,\dot{h}, \tag{7.35}$$

which is called the *equation of free streaming*. With appropriate approximations, the equation (7.35) can be solved directly and used to close the hierarchy of equations.

The value of the brightness function $\delta^{(r)}$ at time t_0 is connected with the fluctuations observed today in the temperature of the

CMB. In fact, the autocovariance function $C(\vartheta)$ of the sky at the present time is just

$$C(\vartheta) = \frac{1}{2\pi^2}\int_0^\infty \sum_l (2l+1)\left(\frac{\sigma_l(k,t_0)}{4}\right)^2 P_l(\cos\vartheta)k^2\mathrm{d}k, \quad (7.36)$$

where the integral takes the distribution from Fourier space back to real space and the divisor of 4 is due to the fact that $\delta_r = 4\Delta T/T$.

Complicated though this procedure may appear, it is still an oversimplification of the calculation necessary for relevant cosmological models: one really needs to treat several different matter components using distribution functions rather than the single fluid matter component presented here; relativistic effects for curvature need also to be included (these calculations are in flat space); there are also some detailed technical issues connected with the choice of gauge in which to perform the calculations (e.g. Bardeen 1980; Panek 1986; Stoeger *et al.* 1995b). The equations presented here are therefore merely intended to introduce the reader to the basic ideas, rather than to present the definitive theoretical predictions to be compared with observations. For examples of more technical references on the temperature fluctuations predicted in various detailed scenarios, including more information on the numerical computations, see Bond & Efstathiou (1984, 1987); Vittorio & Silk (1984); Efstathiou & Bond (1986, 1987); White, Scott & Silk (1994); Hu & Sugiyama (1995).

7.3.3 Intermediate scales: the Doppler peaks

An example of a numerical computation of the C_l over the range of interest here is given in Figure 7.1 for variations on the standard CDM model. One first notices a steep increase in the angular power spectrum for $l \sim 100$ to 200. This angular scale corresponds to the horizon scale at $z_{\rm rec}$. The shape of the spectrum beyond this peak is complicated and depends on the relative contribution of baryons and dark matter; notice how the small 'bumps' change position as Ω_b and Ω_0 are changed.

Although these results are computed numerically, it is important to understand the physical origin of the features of the resulting C_l at least qualitatively. The large peak around the horizon scale is usually interpreted as being due to velocity perturbations on the last scattering surface, as suggested by equation (7.24). The features at higher l are connected with a phenomenon called *Sakharov oscillations*. Basically what happens is that perturbations inside the horizon on these angular scales oscillate as acoustic standing waves with a particular phase relation between density and velocity perturbations. These oscillations can be seen in Figure 7.1. After recombination, when pressure forces become negligible, these waves are left with phases which depend on their wavelength and on the time they started oscillating. Both the photon temperature fluctuations and the velocity perturbations are therefore functions of wavelength (both contribute to $\Delta T/T$ in this regime) and this manifests itself as an almost periodic behaviour of C_l. The use of the term 'Doppler peak' to describe only the first maximum of these oscillations is misleading because it is actually just the first (and largest amplitude) such oscillation. Although velocities are undoubtedly important in the generation of this feature, it is wrong to suggest that the physical origin of the first peak in the angular power spectrum is qualitatively different from the others. It is important to point out that, although these oscillatory features are potentially a very sensitive diagnostic of the perturbations generating the CMB anisotropy, it will indeed be very difficult to resolve them: most experiments are sensitive to a range of l which is much broader than the oscillations themselves. For more detailed discussions of the physical origin of these peaks, see White, Scott & Silk (1994); Hu & Sugiyama (1995); Scott, Silk & White (1995); Hu, Sugiyama & Silk (1996).

In terms of the spherical harmonic expansion of the equations of kinetic theory, the origin of the Doppler peak is straightforward to see. The monopole term σ_0 is essentially the perturbation to the radiation density. Perturbation modes with a wave-number k that makes σ_0 an extremum on the last scattering surface are the ones that generate the Doppler peak. Looking at the coupling between these equations demonstrates that the Doppler peak is not simply caused by velocities through the terms in σ_1 but by perturbations generated by all orders of the hierarchy.

An important complication of these discussions is the relatively slow rate of recombination. One effect of this is that the optical depth to the last scattering surface can be quite large, and small-scale features can be smoothed out. For example, in the context of the standard theory of recombination, the last scattering surface can have an effective 'width' up to $\Delta z \simeq 400$ which corresponds to a proper distance now of $\Delta L \simeq 40h^{-1}$ Mpc, and to an angular scale $\simeq 20$ arcminutes. The finite thickness of the last scattering surface can mask anisotropies on scales less than ΔL in the same way that a thick piece of glass prevents one from seeing small-scale features through it. This causes a damping of the contribution at high l and thus a considerable reduction in the $\Delta T/T$ relative to the photon temperature fluctuations.

High angular frequency fluctuations are also quite sensitive to the possibility that the universe might have been reionised at some epoch. As we argued in Chapter 5, we know that the intergalactic medium now is almost completely ionised. If this happened early enough, it could smear out the fluctuations on scales less than a few degrees, rather than the few arcminutes for standard recombination, the case shown in Figure 7.1. Some non-standard cosmologies involve such a late recombination, so that Δz might be much larger. The minimum redshift allowable is, however, $z \simeq 30$ because an optical depth $\tau \simeq 1$ requires enough electrons (and therefore baryons) to do the scattering; a value of $z < 30$ would be incompatible with $\Omega_b < 0.1$; we discussed this in Chapter 5. In any case, if some physical process caused the universe to be reheated after $t_{\rm rec}$ then it might smooth out anisotropy on scales less than the horizon scale at the time when the reionisation occurred. The angular scale corresponding to the particle horizon at z is of the order of $(\Omega/z)^{1/2}$, so late reionisation at $z \simeq 30$ could smooth out structure on scales of $10°$ or less, but not scales larger than this. If this indeed occurred, one might expect to see a significant anisotropy on a smaller angular scale, generated by secondary effects (e.g. Ostriker & Vishniac 1986).

The message one should take from these theoretical considerations is that temperature fluctuations on intermediate and small angular scales are much more model-dependent than those on larger scales. In principle, however, they enable one to probe quite detailed aspects of the physics going on at $t_{\rm rec}$ and are quite sensitive to parameters which are otherwise hard to estimate. In particular, the position of the first 'Doppler' peak in the angular spectrum is particularly important as a potential diagnostic of Ω_0, as long as it has not been obliterated by reionisation. Its position depends relatively directly on the angular scale subtended by the horizon at last scattering and therefore, through equation (7.22), has a simple dependence on Ω_0 as shown in Figure 7.1. Furthermore, the height of the peak depends strongly on Ω_Λ, so there is a possibility of using this as a test of flat models with low Ω_0 and a cosmological constant term. The position of the subsidiary peaks depend on Ω_b, but these will probably be more difficult to measure. In addition, tensor gravitational wave modes do not produce Doppler motions and their contribution to C_l should therefore be small for high l. One therefore has the hope of using a combination of large and small-scale results, together with the detailed shape of an experimentally measured C_l, to constrain the gravitational wave contribution relatively strongly, at least within the framework of a particular model of the dark matter content. The shape of the spectrum should also permit the possible reionsation of the universe to be constrained directly.

7.4 Observational status

After this somewhat lengthy (but nevertheless simplified) treatment of the theory of CMB anisotropies, it is now pertinent to discuss the experimental results available at the present time, and indicate the likely prospects for the future.

7.4.1 COBE and all that

It is appropriate to make some more detailed remarks about the observational detection of anisotropies generated by the Sachs–Wolfe effect. Two experiments, COBE (Smoot *et al.* 1992) and Tenerife (Hancock *et al.* 1993) have announced firm detections of anisotropy at levels consistent with each other in the region of the angular power spectrum, which is expected to be dominated by the Sachs–Wolfe effect.

Such is the importance of the COBE discovery that it is worth describing the experiment in a little detail. The COBE satellite actually carried several experiments on it when it was launched in 1989. The anisotropy experiment, called the DMR, yielded a positive detection of anisotropy after one year of observations. The advantage of going into space was to cut down on atmospheric thermal emission and also to allow coverage of as much of the sky as possible (ground-based observations are severely limited in this respect). The orbit and inclination of the satellite is controlled so as to avoid contamination by reflected radiation from the Earth and Moon. Needless to say, the instrument never points at the Sun. The DMR detector consists of two horns at an angle of 60°; a radiometer measures the difference in temperature between these two horns. The radiometer has two channels (called A and B) at each of three frequencies: 31.5, 53 and 90 GHz, respectively. These frequencies were chosen carefully: a true CMB signal should be thermal and therefore have the same temperature at each frequency; various sources of galactic emission, such as dust and synchrotron radiation, have an effective antenna temperature which is frequency-dependent. Combining the three frequencies therefore allows one to subtract a reasonable model of the contribution to the observed signal which is due to galactic sources. The purpose of the two channels is to allow a subtraction of the thermal noise in the DMR receiver. Assuming the sky signal and DMR instrument noise are statistically independent, the net temperature variance observed is

$$\sigma_{\rm obs}^2 = \sigma_{\rm sky}^2 + \sigma_{\rm DMR}^2. \tag{7.37}$$

Adding together the input from the two channels and dividing by 2 gives an estimate of $\sigma_{\rm obs}^2$; subtracting them and dividing by 2 yields an estimate of $\sigma_{\rm DMR}^2$, assuming that the noise signals on the two channels are statistically independent. Taking these two together, one can therefore obtain an estimate of the rms sky fluctuation. The first COBE announcement in 1992 gave $\sigma_{\rm sky} = 30 \pm 5\ \mu$K, after the data had been smoothed on a scale of 10°.

In principle the set of 60° temperature differences from COBE can be solved as a large set of simultaneous equations to produce a map of the sky signal. The COBE team actually produced such a map using the first year of data from the DMR experiment. It is important to stress, however, that, because the sky variance is of the same order as the DMR variance, it is not correct to claim that any features seen in the map necessarily correspond to structures on the sky. Only when the signal-to-noise ratio is much larger than unity can one pick out true sky features with any confidence. The first year results should therefore be treated only as a statistical detection.

The value of $\langle a_{lm}^2 \rangle^{1/2}$ obtained by COBE is of the order of 5×10^{-6}. This can also be expressed in terms of the quantity $Q_{\rm rms}$, which is defined by

$$Q_{\rm rms}^2 = \frac{T_0^2}{4\pi} \sum_m \langle |a_{2m}|^2 \rangle = \frac{5T_0^2}{4\pi} \langle |a_{2m}|^2 \rangle \simeq (17\ \mu{\rm K})^2. \qquad (7.38)$$

Translated into a value of σ_8(mass) using (7.19) with $n = 1$ and a standard CDM transfer function, this suggests a value of $b \simeq 1$, unless gravitational waves contribute a large part of the temperature anisotropy. In other words the mass fluctuations need to be unbiased. The problem with this for the standard CDM model is that one cannot simultaneously match both the present-day galaxy clustering amplitude and the typical peculiar velocities of galaxies, unless $\Omega_0 h \simeq 0.25$. It would thus appear that CDM is excluded if $\Omega_0 = 1$.

We should say that normalising everything to the quadrupole in this way is not a very good way to use the COBE data, which actually constitute a map of nearly the whole sky with resolution of about 10°. The rms temperature anisotropy obtained from the whole map is of the order of 1.1×10^{-5}. (Both this value and the quadrupole value are likely to change as more data from this experiment are analysed.) The quadrupole mode is actually not as well determined as the C_l for higher l, so a better procedure is to fit to all the available data with a convolution of the expected C_l for some amplitude with the experimental beam response and then determine the best fitting amplitude for the data. The results of more sophisticated data analysis like this are, however, in reasonable agreement with the simpler method mentioned above.

The statistical errors associated with the amplitude and spectrum of CMB fluctuations detected by COBE have changed significantly since the initial announcement (Smoot *et al.* 1992) as a result of different statistical analyses (Smoot *et al.* 1994; Bunn, Scott & White 1995; White & Bunn 1995) and as more data have been gathered (Bennett *et al.* 1996). There is also the possibility that the power-spectrum need not be of the usual Harrison–Zel'dovich form, and that a substantial part of the COBE signal could be associated with gravitational waves rather than density perturbations, so the implications for Ω_0 are not completely clear. What one can say, however, is that models can be found that match the available clustering data with the normalisation required by COBE. Of course, it may be the assumption that the universe contains a dominant contribution of cold dark matter that is wrong, but if that hypothesis is correct then one should favour an open universe. This is also favoured by independent arguments from large-scale structure, which we discussed in Chapter 6.

7.4.2 The search for the Doppler peak

The potential power of the intermediate scale anisotropy pattern as a test of detailed cosmological models is of such importance that many experiments are underway to look for fluctuations on scales from a few degrees (e.g. SP91) down to a few arcminutes (OVRO); see the right-hand panel of Figure 7.1. Many other experiments (ULISSE, Saskatoon, Python, MSAM I and MSAM II to name but a few) are also underway on these scales; some of these are shown in the figure, but we shall not discuss these experimental results in detail, only making a few comments about the problems of interpreting their output.

The fluctuations one is looking for generally have an amplitude of the order of 10^{-5}. One is therefore looking for a signal of amplitude around 30 μK in a background temperature of around 3 K. One's observational apparatus, even with the aid of sophisticated cooling equipment, will generally have a temperature much higher than 3 K and one must therefore look for a tiny variation in temperature on the sky against a much higher thermal noise background in the instrument. From the ground, one also has the problem that the sky is a source of thermal emission at microwave frequencies. Noise of these two kinds is usually dealt with by integrating for a very long time (thermal noise decreases as $\sqrt{t}$, where t is the integration time) and using some kind of beam switching design in which one measures not ΔT at individual places but temperature differences at a fixed angular separation (double-beam switching) or alternate differences between a central point and two adjacent points (triple-beam switching). Recovering the ΔT at any individual point (i.e. to produce a map of the sky) from these types of observations is therefore not trivial. Moreover, any radio telescope capable of observing the microwave sky will have a finite beamwidth and will therefore not observe the temperature point by point but would instead produce a picture of the sky convolved with some smoothing function, perhaps a Gaussian:

$$F(\vartheta) = \frac{1}{2\pi\vartheta_f^2} \exp\left(-\frac{\vartheta^2}{2\vartheta_f^2}\right). \tag{7.39}$$

It is generally more convenient to work in terms of l than in terms of ϑ so one expresses the response of the instrument as F_l; the relationship between F_l and $F(\vartheta)$ is the same as between C_l and $C(\vartheta)$ given by equation (7.10). In the case of (7.39), for example, we get

$$F_l = \exp\left[-\left(l + \frac{1}{2}\right)^2 \frac{\vartheta_f^2}{2}\right]. \tag{7.40}$$

The observed (smoothed) temperature autocovariance function can then be written

$$C(\vartheta; \vartheta_f) = \frac{1}{4\pi} \sum_{l=2}^{\infty} (2l+1) F_l C_l P_l(\cos\vartheta). \tag{7.41}$$

One must also allow for the effect of beam switching upon the measured temperature fluctuations. We shall here just illustrate the effect on the mean square temperature fluctuation. For a single beam experiment this is just

$$\left\langle \left(\frac{\Delta T}{T}\right)^2 \right\rangle = \frac{1}{4\pi} \sum (2l+1) C_l F_l = C(0; \vartheta_f), \tag{7.42}$$

while for a double-beam experiment, where each beam has a width ϑ_f and the *beam throw*, i.e. the angular separation of the two beams, is α, we have

$$\left\langle \left(\frac{\Delta T}{T}\right)^2 \right\rangle = \left\langle \frac{(T_1 - T_2)^2}{T^2} \right\rangle = 2\left[C(0; \vartheta_f) - C(\alpha; \vartheta_f)\right]. \tag{7.43}$$

The function F_l provides the best way to describe the response of any particular experiment. Of course, different experiments are designed to respond to different angular scales or different ranges of l. For example, the COBE DMR experiment we discussed above has a beam-switching configuration with a beam width of a few degrees and a beam throw of around 60°; this experiment is sensitive to relatively small l. Beam-switching experiments, either on the ground or balloon borne probe the interesting intermediate range around 1°. Single-dish ground-based experiments, such as those at Owens Valley Radio Observatory (OVRO), operate at the other end of the angular spectrum and can be sensitive to l modes of the order of several thousand. In the right-hand panel of Figure 7.1 we show some reported detections: the horizontal extent of the error bars indicates the range of l which are expected to contribute significantly to the measured anisotropy, i.e. they measure the rough width of F_l for a particular experiment.

Several different experiments have reported detections, but these are often inconsistent with other experiments on the same angular scale; we shall refrain from giving detailed numerical values here. Sometimes even the same experiment produces detections in different areas of the sky which are incompatible with each other! The problem with these experiments, which are all either balloon or ground-based, is twofold. Firstly, they usually probe a relatively small part of the sky and the signal they see may not be representative of the whole sky, i.e. they are dominated by *sample variance*. The second problem is that they generally do not have the ability to remove point sources (because of the smaller beam) or non-thermal emission (because of the smaller number of frequency channels) as effectively as COBE. While such observational programmes are being pursued with great vigour, they have yet to produce really firm constraints on the form of dark matter or on the scalar to tensor ratio.

7.4.3 Prospects for the future

Although we have been relatively downbeat about the present experimental situation for intermediate and small-scale experiments, because of experimental uncertainties and incomplete sky coverage, there is plenty of cause for optimism that the angular power spectrum can be measured quite accurately in the near future. Two satellite projects, MAP and COMBRAS/SAMBA – the former based at NASA in the United States and the latter a European project planned by ESA, should be able to construct complete sky maps with sufficient sensitivity and angular resolution to trace the first few peaks of C_l. Most cosmologists would agree that, if all goes to plan, these missions, which will fly early in the 2000s, should settle not only the issue of Ω_0 but also, through the detailed shape the temperature anisotropy, yield information on the relative amplitudes of scalar and tensor modes, the size of the cosmological constant and the baryon fraction. Although the predictions are to some extent model dependent, they are based on relatively well-understood physics and provide probably the most direct means of testing cosmological models. There is, of course, always the possibility that the C_l determined from these experiments will be completely unlike any of the model predictions, in which case the implications for Ω_0 will be unclear. But even this would be an important step forward for cosmology, as it would still give a clear indication of the behaviour of the universe in the early stages of its evolution which could then be used to build alternative models.

7.5 Summary

The main conclusions to be drawn from this chapter are as follows.

- The theory of the origin and evolution of CMB temperature anisotropies has been the subject of much detailed work. The machinery for calculating predictions is now sufficiently well-developed that the detailed temperature pattern produced in a whole variety of different scenarios can be calculated straightforwardly and accurately.
- Contrary to the general folklore, the COBE detection of large-scale CMB anisotropies does not rule out low density universe models, but neither does it 'prove' inflation. It does provide a long-wavelength normalisation for the power spectrum of initial density perturbations which is rather insensitive to the precise value of Ω_0 in open universes.
- Following on from the previous item, we stress that there are low density models which fit *all* the available data on CMB fluctuations and provide satisfactory models of structure formation.
- A robust potential indicator of the value of Ω_0 is the position of the so-called Doppler peak in the angular power spectrum. Present experimental results in the vicinity of this peak are inconclusive, but there are two satellite experiments in the pipeline

which should resolve this issue within a few years and so yield a very good indication of the value of Ω_0.

Finally we emphasise that the COBE results directly refer to the period from decoupling to the present day, and only indirectly to processes prior to decoupling, such as inflation (if it occurred). It is important in analysis of the results to separate out these two aspects; the relation of CBR anisotropies to fluctuations at last scattering, and the prior source of those fluctuations (which might be inflation, but then again might not). Given such a clear analysis, the CBR anisotropy data is one of the most important handles we will have in the future on the origin of large-scale structure and on the value of Ω_0.

8
More realistic universe models

Underlying what has been said above are some theoretical issues that remain unresolved, despite their importance for understanding the geometry of realistic universe models. The central feature is that the models on which the analysis above is based are the standard homogeneous and isotropic Friedman–Robertson–Walker (FRW) universe models; but the actual universe – from which our observations derive – is in fact inhomogeneous on small and intermediate scales†.

8.1 Lumpy universe models

A series of related problems arise. The first point is that most methods of estimating Ω_0 cannot determine the background model without simultaneously solving for its perturbations (for example, estimating velocity flows; calculating lensing effects; considering the effects of inhomogeneities on the CMB; and so on). Furthermore, there is a feedback from these inhomogeneities to the background model dynamics (and so to the equations whereby one estimates Ω_0). But that raises the issue of gauge freedom in describing such perturbations, and how one chooses the background model when faced with a 'lumpy' real universe, which is the second main point.

The gauge issue has been intensively studied in recent years by Bardeen (1980), Ellis & Bruni (1989) and many others; see, e.g., Bruni, Dunsby & Ellis (1992) for a summary and references.

† We do not consider here the intriguing possibility that the universe is not like an FRW universe in the large, because the almost-EGS analysis (Stoeger *et al.* 1995a) provides a reasonable basis for proceeding on the standard assumption that it is homogeneous on the largest scales. We simply note here that this is not the only possibility that is compatible with observations; see Ellis (1984a) and Wainwright & Ellis (1996).

The question here relates to the task of finding a description of inhomogeneities that does not depend on the way the background model is mapped to the real universe. The converse problem of the optimal way to fit a background FRW model to the real lumpy universe (Ellis & Stoeger 1987) has been the subject of much less thought. But this is the basis of the definition of Ω_0 – which of necessity refers to an idealised background model rather than the real lumpy universe, for only by introducing such a background model can we get a unique answer for Ω_0 at any cosmic epoch. This is usually done in an *ad hoc* way; but different ways of doing this fitting will result in different estimates of Ω_0. In particular, in the real universe, the value of Ω_0 as inferred from any finite-size averaging volume varies with scale, from place to place and with time‡. Hence an important feature is that different ways of estimating the density will by their nature select different scales of averaging, or will give different answers at different length scales; so one has to consider how to relate the optimal background model to the real universe on different scales. This is the central issue that needs clarity.

Thirdly, the lumpy nature of the real universe not only affects the dynamics of the smooth background model, through a contribution to the effective stress-energy density tensor at large scales when smoothed out from its value at small scales, but also affects the way we see that universe, by perturbing null geodesics. Hence it affects the analysis of observational relations used to determine Ω_0.

All these effects need to be taken into account if we are to be on secure foundations in our estimates of Ω_0. We consider them now, starting off with the central issue of fitting, then looking at effects of lumpiness on dynamics and on observations, and concluding with comments on the issue of averaging scales and determination of Ω_0.

8.2 The fitting problem and integral constraints

The question is how to fit an FRW background model to the lumpy reality of the universe. This procedure *defines* the value of Ω_0 we

‡ For example, in the room where you are reading this, its value is about 10^{+30} (averaged over the scale of the room)!

are searching for, for the real value of that quantity varies spatially because of the inhomogeneity of the real universe. When we quote a global value, we have in mind some smoothed out background model that represents the nature of the real universe considered in the large. The associated averaging procedure has been little studied; it is equivalent to defining a best-fitting procedure for fitting a smooth background model to a lumpy universe model. The way this is done can have significant implications, as is made clear for example by the Traschen integral constraints (Traschen 1984). These can be approached in two ways.

Consider an almost homogeneous universe model in which inhomogeneities form by local physical processes. Then local energy and momentum conservation imply the existence of conserved quantities that, physically speaking, correspond to conservation of monopole and dipole terms (because energy and momentum are conserved). This implies integral relations that must be satisfied in the perturbed model, if it is to be consistent with formation of the inhomogeneities by re-arrangement of matter, starting from a completely smooth background model.

In models where we take these integral constraints into account, the various ways of estimating Ω will be affected (relative to an analysis where we do not take them into account). In particular, the Sachs–Wolfe effect is reduced relative to models where the effect is ignored (Traschen 1984; Traschen & Eardley 1986); consequently estimates of Ω_0 based on CMB fluctuations will be affected. Now it can be claimed (Traschen & Eardley 1986) that this argument does not apply to matter perturbations created by quantum fluctuations during the inflationary era in inflationary universe models, because of their non-local size (in terms of today's scales). However, there is another way of looking at these constraints that does apply in this case too: they can be thought of as the fitting conditions required to be satisfied if the FL background model we choose corresponds correctly to the real lumpy universe (Ellis & Jaklitsch 1989).

The point can be made more concrete by remarking that we might start off with a uniform universe model A with background density Ω_A, and then model high-density regions by adding overdensities at various places to model clusters of galaxies there, resulting in a non-uniform model B. Now the obvious but important consequence of this is that the average density Ω_B in B is then greater (in any finite-sized averaging volume containing the over-density) than the average density Ω_A in A. Thus if employing model B, we must remember to use for the background model the higher value Ω_B rather than the starting value Ω_A – which means changing to a new background model A' with the correct density ($\Omega_{A'} = \Omega_B$), and consequently with different dynamics than those of the model initially contemplated.

If we insist on using the same background density value in the lumpy universe as in the original background model A, we must replace model B by a model C where the high-density regions are surrounded by void regions (that is, regions with density less than the background value Ω_A) so that the average density does indeed work out correctly: $\Omega_C = \Omega_A$. This is essentially the condition expressed by the Traschen scalar constraint equation (Ellis & Jaklitsch 1989). It corresponds to the fact that C can be obtained from A by re-arranging matter in it (which is not true for B).

Whichever approach is adopted, both the predicted Sachs–Wolfe relation and other dynamic relations are different if these constraints are satisfied than if they are not. However, many analyses do not explicitly require that they be satisfied. For example, attempts have been made to estimate Ω_0 by analysing the motion of the Local Group with respect to the Virgo cluster of galaxies (e.g. Sandage & Tamman 1990). Often the method is to model the Virgo cluster as a spherically symmetric overdensity sitting on an FRW background. The resulting model therefore corresponds to a model with a higher total density than the original background model, unless some compensating voids are introduced. These voids would also have a dynamical effect upon the Local Group and therefore alter the conclusions: velocities inside the compensating low density region will be higher, but velocities outside it will be lower. The analysis of velocities will need to take this into account.

For present purposes the point is that if we take this effect into account, the Traschen & Eardley results assist us in sustaining the conclusion reached previously, that there is no contradiction between the observed low anisotropy levels of the CMB and the formation of large scale structures in the available time in low-density universes. The effect certainly needs to be taken into account in any method of estimating Ω_0 from inhomogeneities.

8.3 Dynamical and observational effects of clumpiness

Related to the use of more realistic 'lumpy' models are two further issues of how the lumpiness affects dynamics and observations.

8.3.1 Effects on the field equations

The field equations are correct to a high order of accuracy for the Solar System and for relatively small binary systems, but that does not say that they are correct for application to much larger scale systems, where averaging has to take place to allow their use at scales quite different from those where they are verified. It is entirely plausible that correction terms need to be added to the equations to account for the fact that the metric that is being used in cosmology is in some sense an average metric. Such polarisation effects occur in electrodynamics – surely they should be expected in gravity also? The issue is a fundamental but seldom considered question: *how can a background model be constructed by averaging of an inhomogeneous universe model*? (see e.g. Shirokov & Fisher 1963; Ellis 1984b).

This effect is difficult to examine in the case of a non-linear theory such as general relativity; however, analysis by Buchert & Ehlers (1993) makes clear that it occurs also in Newtonian theory.

Newtonian theory

Using the Newtonian evolution equations a general expansion law has been derived by Buchert & Ehlers (1993), which shows that the Friedmann-type behaviour of the expansion is modified by the presence of inhomogeneities – the averaged rate of squared shear and vorticity and the fluctuation of the expansion rate contribute on average to the global expansion of the universe. This is a scale-dependent effect: on smaller scales the effect is expected to be larger, while on some large scale the standard Friedmann model may describe the average flow correctly.

In detail: let $(\rho, \mathbf{v})$ be any solution to the Euler–Poisson equations for pressureless matter discussed in Chapter 6, and let $\Theta =$

$\nabla\cdot\mathbf{v}$, ω and σ denote the rate of expansion, magnitude of rotation and magnitude of shear, respectively. Let

$$\frac{\mathrm{d}}{\mathrm{d}t} \equiv \frac{\partial}{\partial t} + \mathbf{v}.\nabla \tag{8.1}$$

denote the comoving time derivative, and write

$$\langle f\rangle \equiv V^{-1}\int_{\mathcal{D}} f\mathrm{d}^3x \tag{8.2}$$

for the spatial average of any field f, taken over the time-dependent region $\mathcal{D}(t)$ with volume $V(t) = a^3(t)$ of a part of the fluid specified by a compact initial region $\mathcal{D}_0$. Then, $a(t)$ corresponds to the cosmic expansion function for any fair sample of the universe; $a^3\langle\rho\rangle$ is the (constant) mass of the matter considered. It then follows that $\langle\Theta\rangle = 3\dot{a}/a$, and

$$\frac{\mathrm{d}}{\mathrm{d}t}\langle\Theta\rangle - \left\langle\frac{\mathrm{d}}{\mathrm{d}t}\Theta\right\rangle = \langle\Theta^2\rangle - \langle\Theta\rangle^2. \tag{8.3}$$

These relations imply the averaged Raychaudhuri equation:

$$3\frac{\ddot{a}}{a} + 4\pi G\langle\rho\rangle - \Lambda = \frac{2}{3}\left(\langle\Theta^2\rangle - \langle\Theta\rangle^2\right) + 2\langle\omega^2 - \sigma^2\rangle \tag{8.4}$$

(Buchert & Ehlers 1993). For a Friedman-type model, the right-hand side vanishes, and (8.4) reduces to the Raychaudhuri equation of Newtonian cosmology. The equation shows how the averaged background variables a, $\langle\rho\rangle$, which depend on the chosen averaging domain $\mathcal{D}$, are affected by inhomogeneities on any scale. In the real inhomogeneous universe, the dominant 'disturbance' presumably is the shear, which acts like an enhanced density. Whether this effect is appreciable depends on the value of the ratio $\langle\sigma^2\rangle/2\pi G\langle\rho\rangle$. To estimate it requires a detailed model of the cosmic velocity field.

If the velocity field $\mathbf{v}$ is decomposed into a Hubble flow with expansion rate $H = \dot{a}/a$ and a peculiar velocity field $\mathbf{u}$, (8.4) can be transformed into

$$3\left(\frac{\ddot{a}}{a}\right) + 4\pi G\langle\rho\rangle - \Lambda = a^{-3}\int_{\partial\mathcal{D}}(\mathbf{u}\nabla\cdot\mathbf{u} - \mathbf{u}\cdot\nabla\mathbf{u}).\mathrm{d}\mathbf{S}\,. \tag{8.5}$$

Thus, for a model with periodic fields, as used in all numerical simulations, one can average over the total space and the right-hand side vanishes. More generally, equation (8.5) offers a way to estimate how deviations from the Friedmann equation depend on the averaging region and the peculiar velocity field.

The broad tendency of this effect can be inferred from the above equations. For example, the Hubble constant measured on scales where this effect contributes is larger than the 'true' Hubble value defined on large scales. Accordingly, the age of the universe inferred from small-scale measurements may underestimate the true age, and the density parameter may be overestimated compared to the true value, in accordance with the results of Bildhauer & Futamase (1991); see also Wu *et al.* (1995), Suto *et al.* (1995) and Shi, Widrow & Dursi (1996). This tendency of the effect works in the right direction, as a low Hubble constant helps solve many of the problems which arise in large-scale structure modelling, a lower Ω_m on large scales helps to match the nucleosynthesis value with the baryon abundance measured on smaller scales, and the effect assists with the age problem. However, the quantitative impact is difficult to estimate.

If structures in the universe form in a highly anisotropic way, as is often proposed, then the average shear will contribute with a positive sign to the apparent mass content, thus affecting the global expansion. In principle this extra term could be responsible for at least some of the 'missing mass'. This needs investigation, involving detailed modelling, since it could impact the interpretation of observations and the dark matter problem in general, particularly because structure formation scenarios are normalised on the basis of the standard model.

It is clear the effect takes place; the issue is whether the velocities involved are large enough that it is important.

General relativity

Analysis of the general relativity case, with its non-linearities, is more difficult, particularly because averaging cannot be defined easily in a generic curved space-time. Nevertheless the studies carried out make clear that the effect occurs there too. Generic analyses of the effect have been carried out, as well as studies of how it works in specific geometries, for example Tolman models and perturbed Friedmann models; see for instance Futamase (1989) and Zotov & Stoeger (1992).

The provisional conclusion (Zotov & Stoeger 1995) is that, as in the Newtonian case, small-scale estimates of the the 'Hubble constant' in a complex system will be different than those determined by measurements at a larger scale, and this effect could even possibly significantly affect the age issue issue: Bildhauer & Futamase (1991). Furthermore the way in which this occurs depends on the actual hierarchy of mass clustering in the universe.

8.3.2 Effects on observations

A similar issue arises in analysing observations in an inhomogeneous universe model. The inhomogeneities affect the dynamics of photons in the universe, altering the shear and the convergence of null rays in lumpy universes, and hence giving effective density terms in the null Raychaudhuri equation. This affects the usual cosmological area–distance relation, and so apparent sizes and luminosities (e.g. Bertotti 1966). Consequently observational relations, such as the relation between number count or angular diameters and redshift, are affected, and so could lead to alterations in the estimation of Ω_0 from galactic observations.

This effect can be taken into account in various ways, but the form of the focusing depends strongly on how clumped the matter is. A useful idealisation is the Dyer–Roeder equation; see §4.5.3 of Schneider, Ehlers & Falco (1992). This equation represents a criterion for the largest possible angular diameter distance (for a given redshift) for light bundles which have not yet passed through a caustic. It is accurate for universes 'not too filled by clumps'. A full analysis requires much more complex models. In particular, although the bending of light is very weak, it leads to formation of caustics associated with gravitational lensing (in effect there is a very long lever arm, so although light rays are deflected a very small amount, they soon run into self-intersections – which are non-linear effects). Consequently the areas associated with given solid angles can be much larger than estimated if lensing is not taken into account. This could seriously affect implications of number counts at high redshift.

8.3.3 Implications

Both effects – altered dynamical behaviour and altered geodesic paths in lumpy models – can change the theoretical relations which underlie the standard analyses significantly, in a way that depends strongly on the clustering of matter in the universe. Thus what has been said in this volume about the evidence and Ω_0 has to be regarded as provisional, subject to a resolution of the effect of lumpiness on observations in low density universes. But the same applies equally to the standard analyses that are claimed to lead to the conclusion that $\Omega_0 = 1$ to high accuracy. None of the present analyses have satisfactorily resolved this issue.

8.4 The issue of scale

Finally underlying all this is an important topic almost never explicitly discussed: on what scale is the FRW model employed supposed to describe the universe? What averaging scale are we referring to when we give our value for Ω_0?

It can be suggested that, on referring to the scale on which we observe the largest anisotropic structures, i.e. about $100h^{-1}$ Mpc, the likely scale above which the average model is of Friedman type coincides with the fair sample scale, estimated to be at least $300h^{-1}$ Mpc. For example, Tini Brunozzi *et al.* (1995) show that, with typical assumptions about the form of the power spectrum matter distribution, the convergence of the cosmological dipole (§6.3.4 above) within a volume V is a necessary condition for the scale of large-scale homogeneity to have been reached. However, this is a model-dependent, rather than an observationally proven result. It needs to be treated with caution. Assuming it is correct, how does it relate to measurements of Ω_0?

8.4.1 Local and global values of Ω

It is obvious that some averaging scale S is implied in any measured value for the density. For example, we can use the mass-to-light ratio for particular systems to obtain an effective value of Ω_0, say $\hat{\Omega}_0$ for the scale represented by these systems. The problem is what happens to this inferred $\Omega_0(S)$ as S is taken to larger and larger values. Observations seem to suggest that $\hat{\Omega}_0(S)$ increases from a very small value $\hat{\Omega}_0 \simeq 0.05$ for small, but still cosmological, volumes to larger values (though not necessarily $\hat{\Omega}_0 = 1$) for large volumes.

If one is to accept that $\Omega_0 = 1$ globally, i.e. on scales greater than $300h^{-1}$ Mpc, then measured values $\hat{\Omega}_0(S)$ would have to increase substantially with S from those measured on small scales, until this cosmological smoothing scale S_c at which $\hat{\Omega}_0(S_c) = 1$. But this cannot happen if the averages at smaller scales are taken over 'representative volumes': for there must be the same amount of mass in a set of small representative volumes as in a set of larger volumes with the same total volume (if the density is to add up correctly). To get such an increase one therefore requires either that the smaller volumes be unrepresentative in some way (with a lower density than a truly representative volume would contain); or that the measure used only takes into account mass distributions that are inhomogeneous on the scale S (as indeed is the case with dynamic estimates; see Chapter 5), so that the estimate is a partial rather than complete estimate of the matter in the volume. But then this should be very clearly stated as an essential part of the analysis.

8.4.2 Selection effects related to scale

Of course, complete samples probing small scale structure will, by practical necessity, be localised in a relatively small volume and therefore dominated by a small number of voids or clusters (perhaps only one) that are contained in them. If the selection of the sample imposes some bias in the form of structures it contains then there is no reason why the sample need be representative. The fact that one normally tries to measure the mass by using galaxies automatically imposes a bias. Dynamical studies of high-density regions do not sample the voids and will therefore produce a systematic over-estimate of the global density, unless the largest contribution to Ω_0 is a completely smooth dark matter component (which the dynamic studies will not measure).

Even all-sky surveys centred on the observer can be biased. The local supercluster dominates within a nearby sample, so samples must be large enough to encompass many representative superclusters (and their compensating voids) before one can probe the cosmological average.

8.4.3 Estimating the packing fraction

If density measures are made on a 'small' scale, one can estimate the contribution of density inhomogeneities measured at that scale to the global value of Ω_0 *provided* one knows the *packing fraction*: i.e. what fraction of the universe as a whole is occupied by such overdense regions. But one cannot estimate that fraction unless one can estimate accurately the fraction of the universe occupied by voids of *all* sizes greater than or equal to the scale S concerned. And there lies the problem.

This reasoning does suggest that it may be possible to reconcile a critical-density universe with the small value measured locally. That is not the only interpretation, however. It is perfectly possible, if there is sufficient large-scale structure, that measured values of Ω_0 could continue to grow up to and including clusters of galaxies, say reaching a value of 0.1 or 0.3, but it could then become lower again as larger scales are considered, even reaching the low value of 0.04 on the largest scales (as the effect of large voids is taken into account).

One could even take this argument as far as to suggest that we live in a strictly hierarchical universe (e.g. de Vaucouleurs 1961) where there is no non-zero limit to Ω_0 as we look at larger and larger averaging volumes. In this case measured values of $\Omega_0(S)$ decrease toward zero as S increases at the largest scales, where the density estimates measure the effects of hierarchical voids as well as hierarchical density concentrations, and so the measures become more and more representative of the true density of matter in the universe. The various suggestions that the universe might have a fractal structure could support this viewpoint, if taken seriously†. We are not supporting this view as realistically describing the universe, but are pointing out that it remains a possibility which should be considered.

Horizons and scales beyond the horizon

A warning is in order here: the largest scale on which we can *in principle* measure densities is the horizon scale, because we simply do not have access to information from further regions. Thus the

† As do all those physical analyses that assume that space-time is asymptotically flat – for example, almost all black hole studies!

attempt to observationally determine a global (i.e. larger than horizon-size) density value is in vain. We can assume what we like about larger scales without fear of contradiction; the universe may be uniform on larger scales (with the same properties as inside the horizon), as was almost always assumed before the advent of inflation; or it may be quite different on the largest scales (as is assumed in chaotic inflationary models). There is no way whatever of measuring the average density of matter on the very largest scales in such models, and indeed such an average may not exist.

Hence we should be clear that the best that observational efforts can do is determine a value of Ω_0 for the best fit model to the observable region of the universe, which is roughly of Hubble scale. Statements that purport to give the situation on larger scales are not observationally based; rather they represent the author's metaphysical position.

8.5 Taking inhomogeneity seriously

These considerations lead us to stress the point that there is an underlying assumption throughout this book that there is a good limit to $\hat{\Omega}_0$ if we use large enough averaging volumes; and we reach that limit on length scales greater than or equal to S_c. The point to be made here is that the dependence of measurements on the size and location of the averaging volume should be made explicit in our discussions, so that the way this limit takes place can be carefully examined. Then in considering the implications of each measurement, we can explicitly consider: Which scale density is being determined by this measurement? How will this relate to the 'global' value of Ω_0?

We regard the question of modelling inhomogeneities as an unsettled issue, which analyses of both high and low density universes should take seriously. We do not estimate these effects in detail here; indeed the needed analyses have not yet all been carried out. Rather we comment that the reader should be aware of these issues, and ultimately only believe estimates that attempt this analysis in a satisfactory way. It is one of the more important – and difficult – theoretical issues that needs clarification.

9
What is the verdict?

What we have done in this book is to look at those aspects of cosmology in which the density of the universe plays a role as either a prediction or a parameter of a model, and compared them with the data. It is now our task to weigh up the arguments we have described, and try to make a reasoned assessment of their implications. This will be done by a forensic approach: some of the evidence is quite reliable, but some of it is purely circumstantial, some unreliable, and some contradictory. In view of this we shall not adopt the criterion of proof that applies in the criminal court ('beyond all reasonable doubt'). Rather we look at the 'balance of the evidence', as in a civil case. Doing this, we believe that despite some counter-indications, it is possible to discern a strong case for a low-density universe having negatively curved spatial sections, i.e. to conclude on the balance of evidence and argument that we live in an *open* universe.

The starting point for this conclusion is that, as should be clear on reviewing the various considerations laid out in the previous chapters, there is no convincing observational case to be made for a critical density universe: the strongest motivation for the supposition that Ω_0 is very close to unity comes from theory rather than observation. We consider the theoretical arguments most commonly advanced, which rely on presuppositions about physics which are not amenable to direct test†, to be inconclusive for reasons discussed at length in Chapter 2. Moreover, whatever the status of these theoretical arguments, our position (argued at length in Chapter 1) is that this question will have to be decided empirically in the end. We shall not therefore discuss theoretical

† Except for the possible laboratory detection of dark matter: but this has not happened.

issues further‡ during this summing up.

However, the lack of a case in favour of $\Omega_0 = 1$ does not constitute a case in favour of any particular alternative model. Reviewing the evidence, we believe further that at present there is a strong, though admittedly not watertight, case for a standard FRW universe with

$$\Omega_0 \simeq 0.2, \quad \Lambda = 0, \quad \kappa = -1. \tag{9.1}$$

The reasons we believe this will become clear as our argument develops. If this is correct, we deduce that the universe will expand forever: it is not in one of a series of ever-repeating cycles, as suggested for example by Tolman (1934). Rather it will suffer the 'heat death' of an everlasting cooling down and accompanying decay of cosmic matter (Islam 1983). Furthermore, if it has the natural spatial topology†, and the universe outside our past light cone is similar to the region accessible to our observations, then its spatial sections are infinite and there is an infinite amount of matter in the universe. In that case, the finite nature of the genetic code implies it is highly probable each of us is accompanied in the universe by an infinite number of identical twins (Ellis and Brundrit 1979). The alternative is that conditions outside our past light cone are quite unlike those within it (cf. Ellis 1975): we live in a low-density bubble in a vastly larger universe that is in general quite unlike the region we can see (as in chaotic inflationary models). Observations will never be able to distinguish between these two possibilities. However, we can achieve reasonable certainty within our past light cone. If we are right in our assessment, then we can conclude that our part of the universe will expand forever, and it will eventually appear as if we live in an island universe (Rothman & Ellis 1987).

‡ We cannot, however, resist one small comment. If indeed $\Omega \simeq 1$ to very high accuracy, say $\Omega_0 = 1 \pm 10^{-4}$, then we will never be able to determine if the universe is in fact open ($\Omega < 1$) or closed ($\Omega > 1$). It is observationally far preferable if we can determine this property, which is of considerable theoretical interest (see the discussion in Chapter 1).

† E.g. we do not live in a 'small universe' we can see right round, cf. Ellis and Schreiber (1986); recent analysis of the CBR data suggests this is not the case, see De Oliveira Costa & Smoot (1995)

9.1 The evidence

We now briefly run through the various arguments again to establish our view that there is a consistent case to be made for a low density universe with $\Omega_0 \simeq 0.2$, where for the sake of this argument $\Omega \simeq 0.2$ means, roughly speaking, $0.1 < \Omega_0 < 0.3$. Though not supported by all the evidence we have discussed, we believe that this model has a compelling degree of consistency with a number of independent arguments and, moreover, that a value of $\Omega_0 \simeq 0.2$ agrees with the most reliable evidence.

In Chapter 3, we argued that presently favoured values of the Hubble parameter are in conflict with estimated ages of globular clusters if $\Omega_0 = 1$. Although there is still some uncertainty in stellar evolution theory, it would be surprising if the ages of the globular clusters could be reconciled with an Einstein–de Sitter model, particularly if the ages and h turn out to be at the upper end of their allowed error bars. Reconciliation† is not so problematic in a universe with $\Omega_0 \simeq 0.2$.

We also argued that the classical cosmological tests were generally dominated by evolutionary effects and therefore did not put clear constraints on cosmological parameters such as Ω_0. We did note, however, that the test which one might have most reason to take as reliable – the use of supernovae as standard candles – along with a less direct argument to do with gravitational lensing probabilities, seem to exclude the presence of a cosmological constant large enough to reconcile $\Omega_0 = 0.2$ with flat spatial sections.

The abundances of light atomic nuclei were discussed in Chapter 4. We demonstrated that there is an impressive agreement between the abundances of helium-4, deuterium, helium-3 and, perhaps, lithium-7 with the predictions of primordial nucleosynthesis, but only if the contribution of baryonic material to the critical density is relatively small: $\Omega_b < 0.10$ if $h = 0.5$ and we allow the most conservative interpretation of the errors in element abundances. There is reason to believe that $\Omega_b < 0.03$ (for $h = 0.5$) if one takes the recent high values of the deuterium abundance seriously. Both constraints get tighter if one adopts a higher

† Lowering the density from $\Omega_0 = 1$ to $\Omega_0 = 0.2$ would only gain about 20 per cent in terms of the predicted age of the universe, but this is now less than the published uncertainty in the ages.

value of h as the measurements discussed in Chapter 3 seem to indicate. If there is other, independent, evidence that the total density is significantly higher than this (and we believe there is) then the extra matter has to be non-baryonic. There remains a possibility that primordial nucleosynthesis might be different in various non-standard scenarios, but the agreement of all the light element abundances seems difficult to achieve in anything much different from the standard model.

We presented a number of astrophysical arguments in Chapter 5. Aside from detailed constraints on the amount of hot gas or neutral hydrogen, which are no more than consistency checks on Ω_b, the main evidence we presented concerns the dynamical behaviour of galaxies and clusters. This provides evidence for significant quantities of gravitating material over and above what can be observed through the radiation it produces. The dark matter directly inferred to exist in galaxies provides a lower limit on Ω_0 but, because these objects are small compared to cosmological scales, they do not necessarily provide a reliable estimator of the mean density: material may be distributed outside galaxies, and thus escape detection.

In the case of clusters, the dynamical evidence for dark matter is supported by independent observations of gravitational lensing. But even clusters are relatively small-scale structures, so, if one is prepared to extrapolate the apparent trend of increasing M/L with scale out to very large scales indeed, then one can obtain the required value for closure. The data, however, do not require this to be the case; see Figure 9.1, which shows clear evidence that the observed M/L increases from galactic to cluster scales, but no evidence of a further increase thereafter. A global value of $M/L \simeq 300$ corresponding to $\Omega_0 \simeq 0.2$, rather than the $M/L \simeq 1400$ required for $\Omega_0 = 1$, is perfectly compatible with the data. If the higher figure were true, one would have to give a clear argument as to why the smaller-scale measurements gave answers different from the global average (cf. the discussion in Chapter 8). Notice that this density is already high enough to exceed the nucleosynthesis bound discussed above for any reasonable value of the Hubble parameter, and therefore requires the existence of non-baryonic dark matter to make up the difference between Ω_b and Ω_0. Moreover, X-ray data show that the fraction of the total mass of clusters

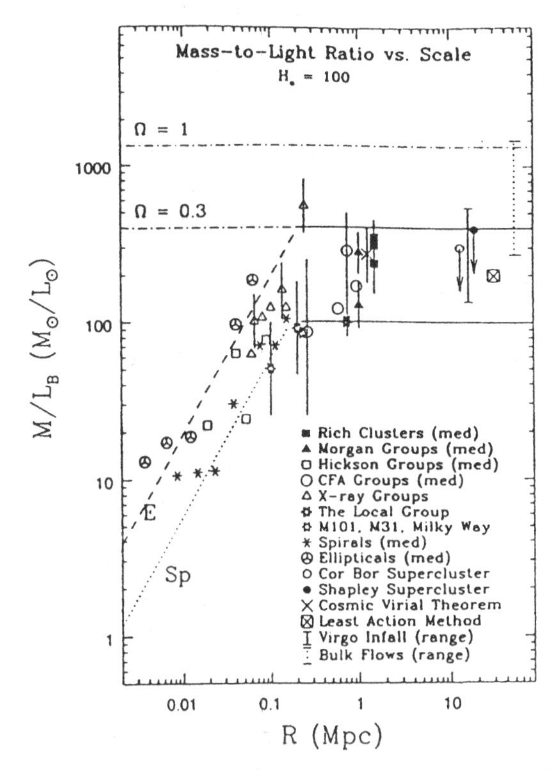

Fig. 9.1 A compilation of estimates of M/L as a function of scale. From Bahcall, Lubin & Doorman (1995).

contained in galaxies and hot gas is only compatible with the nucleosynthesis constraints discussed above if $\Omega_0 \simeq 0.2$. Unless there is something wrong with the theory of cluster formation, then these results argue strongly for a low-density universe.

In Chapter 6, we discussed at some length the question of the origin of galaxies and large-scale structure. Not surprisingly the evidence in this field is highly uncertain and often contradictory. In particular, the theoretical framework within which one interprets large-scale clustering and peculiar motions is highly model-dependent: the mean density of the universe is only one parameter involved in theories of the origin of galaxies and large-scale structure and in many circumstances plays only a minor role in the proceedings. A strict judge might therefore be persuaded to rule that all this evidence should be inadmissible for these reasons. On the other hand, large-scale structure studies have so far provided the only sustainable observational estimates of $\Omega_0 \simeq 1$. For this reason, we shall include the large-scale evidence in these deliberations. In any case, we feel the balance of even this evidence is consistent with a universe having $\Omega_0 \simeq 0.2$. For a start, the behaviour of galaxy clustering statistics is well fitted by a power-spectrum corresponding to a low-density CDM universe, with $\Omega_0 \simeq 0.2$. Such a model, if normalised to COBE (see below), also matches the observed cluster abundances, seems to be indicated by clustering evolution arguments and is consistent with the baryon fraction estimate presented above. Spatial clustering statistics provide the most indirect argument of all, however, because the relationship between galaxies and mass might be very complicated. Use of velocity information is therefore required to put these estimates on a more solid footing.

Among the dynamical arguments, galaxy velocity dispersions on small scales have consistently yielded low values for Ω_0, consistent with the dynamically inferred masses for clusters discussed above. On larger scales, there are basically three types of evidence: (i) dipoles; (ii) peculiar velocity studies; and (iii) redshift-space distortions. The first of these (i) is, at least with currently available data sets, limited by the requirement for dipole convergence with the sample depth. We presented evidence that this has not been achieved in such studies to this date. Peculiar-motion studies (ii) require an estimate of the distance to a galaxy which is independent of its redshift. Again we believe that there is certainly reasonable doubt as to whether the distance estimators are sufficiently reliable. The method of redshift-space distortions (iii) does not require independent distance estimates but, because of its statistical character, does require large samples of galaxy redshifts. We stress, however, that although these methods do in principle probe the dynamics (and hence the total gravitating mass) of large-scale structures, one still needs to make assumptions about the relationship between galaxies and mass. One can only estimate Ω_0 if one accepts that the parameter β (which is what is actually measured) can written $\Omega_0^{0.6}/b$, the numerator coming from the linear theory of gravitational instability, and the denominator coming from the linear bias model. If either of these is not appropriate, then these methods simply do not measure Ω_0. Even if we do accept this interpretation of β, it is interesting to note that (i) and (ii) have both yielded claims of $\beta \simeq 1$, while (iii) has tended to generate $\beta \simeq 0.5$. We believe that if one is going to trust any of these arguments at all then one should trust (iii), as it is more robust to the presence of systematic errors. If one adopts $b = 1$ then $\Omega_0 \simeq 0.3$, consistent with the rough figure we have been proposing†.

We should stress again that these considerations would in any can only yield model-dependent constraints on Ω_0. The entire gravitational instability picture, upon which they are all based, may be completely wrong. At the moment, there is no good reason to believe this, because this picture provides a framework which explains much that we know about galaxy clustering. That does not, however, prove that it is right. It may be that some vital physical ingredient is not included in this theory. For example, it may be that primordial magnetic fields have a significant effect on the process of structure formation (Coles 1992).

The final pieces of evidence we discussed (in Chapter 7) come from the microwave background radiation. We showed that, contrary to some published claims, the COBE data, either on their own or in conjunction with large-scale structure data, do not constrain Ω_0. At the moment, the data from intermediate scale experiments (around 1°), where the sensitivity to Ω_0 is strongest, yield inconclusive results. Prospects for microwave background studies to pin down not just Ω_0, but also Ω_b, h and Λ in the relatively near future are, however, extremely bright particularly now that both ESA and NASA have selected CMB probes to be launched early in the next century.

9.2 The weight of evidence

In the light of this discussion, we believe that the weight of evidence lies towards a value of $\Omega_0 \simeq 0.2$. A density much higher than this, and particularly a value $\Omega_0 = 1$, would require the following

† It is worth remembering that, as we mentioned in Chapter 6, the idea of biasing was introduced into the CDM model in order to reconcile dynamical estimates of the cosmological density yielding $\Omega_0 \simeq 0.2$ with the theoretical preference for $\Omega_0 = 1$. If Ω_0 is less than unity, then there is no need to assume any biasing at all, so it is reasonable to take $b = 1$ in this case.

to be true:

- *either* recent estimates of h are too high *or* the ages of globular clusters have been overestimated (or both);
- *either* the primordial nucleosynthesis limits are wrong *or* the understanding of cluster formation and dynamics is wrong (or both);
- there is a continued increase of observed M/L with scale beyond the scale of galaxy clusters;
- some ingredient other than CDM is involved in large-scale structure formation and evolution;
- galaxy formation is biased;
- peculiar-motion studies must somehow be giving a better estimate of Ω_0 than redshift-space distortions.

By contrast, none of these need be true if $\Omega_0 \simeq 0.2$.

Our argument against the alternative picture with $\Omega_0 \simeq 0.2$ and $\Omega_\Lambda \simeq 0.8$ is less strong. In this scenario the age problem is by-passed, but one requires instead that the QSO lensing constraints and the Type Ia supernovae limits are both incorrect. These could be avoided if, as for example Krauss & Turner (1995) have suggested, we have a model with $\Omega_0 \simeq 0.4$ and $\Omega_\Lambda \simeq 0.6$, but this value of Ω_0 is towards the upper end of the range of directly inferred values, among which we have found consistency with $\Omega_0 \simeq 0.2$. Most other aspects of the behaviour of the flat Λ model are similar to the open version with the same Ω_0. In particular, gravitationally induced large-scale motions are only weakly influenced by Ω_Λ, so such considerations do not discriminate between these alternatives.

Although the Λ model is not ruled out as strongly as $\Omega_0 = 1$ appears to be, we prefer the negatively curved model on the additional grounds that it requires fewer assumptions about unknown physics to fit the data: in many ways the introduction of a cosmological constant of the required magnitude poses more problems than it solves (see Chapter 2).

It is timely at this stage to look back at the classic paper of Gott *et al.* (1974). They weighed up the evidence available at that time from nucleosynthesis, galaxy masses, small-scale dynamics, age considerations and so on, much as we have done in this book, although without the benefit of the considerable amount of data

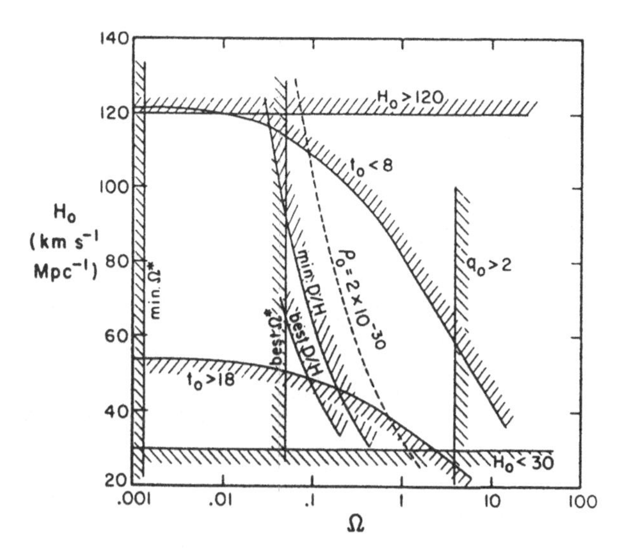

Fig. 9.2 Figure from Gott *et al.* (1974). This shows the constraints on Ω_0 and H_0 placed by age considerations (t_0), the error range of H_0 itself, and considerations from nucleosynthesis (D/H) and mass estimates from galaxies Ω^*. The allowed region corresponds to $\Omega_0 \simeq 0.1$ and $h \simeq 0.6$.

that has become available in more recent years. The conclusion then was that the only value of Ω_0 compatible with all the observations was $\Omega_0 \simeq 0.06 \pm 0.02$. In the 20 years since that paper was written, much more data has been accumulated and some of the constraints have tightened: better knowledge of physical parameters has reduced the allowed range of Ω_b compatible with nucleosynthesis; the evidence for dark matter from galaxies, clusters and large-scale structure arguments has accumulated also, but the idea that some of this may be non-baryonic makes its implications for this diagram less clear. In any case, the value quoted by Gott *et al.* (1974) is only a factor of two or three away from the range of Ω_0 we suggest.

It is enlightening to compare Figure 9.2 with the corresponding diagram† from Krauss & Turner (1995), which shows constraints similar to those we have discussed here. We disagree with the

† See also Ostriker & Steinhardt (1995) for similar arguments.

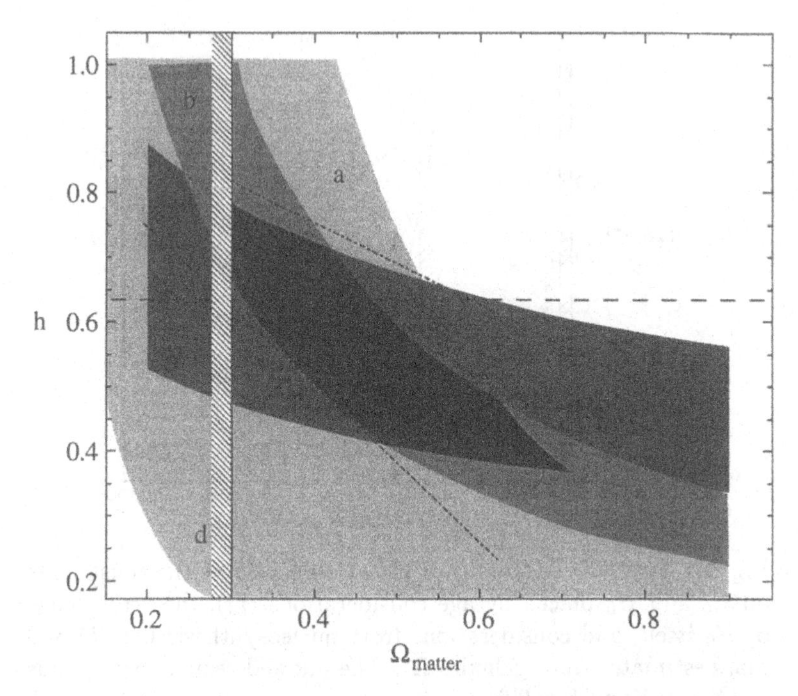

Fig. 9.3 Figure from Krauss & Turner (1995). An updated version of Figure 9.2, this shows the constraints on Ω_0 in matter and h placed by (a) the combination of nucleosynthesis constraints with the cluster baryon fraction, (b) the shape of the clustering power spectrum on large scales, (c) age considerations and (d) large-scale motions. In contrast to the previous figure, this diagram assumes the existence of a cosmological constant to ensure that $\kappa = 0$.

emphasis in Figure 9.3, which gives the same weight to the constraints arising from the large-scale galaxy clustering power spectrum (with all its associated uncertainties and model-dependence) and peculiar-motion studies, as it does to the age constraint and, above all, is predicated on the assumption that $\kappa = 0$. We also believe that the allowed region should extend as far left as $\Omega_0 \simeq 0.2$. Nevertheless, the fundamental point we have made is contained in Figure 9.3: $\Omega_0 = 1$ is ruled out principally by the age argument, but also by a number of supporting considerations.

Assessing this situation overall, we find substantial evidence (including that we regard as more reliable) in favour of our hypothesis, and considerable problems with both major competing proposals. Hence we arrive at our conclusion (9.1).

9.3 Criteria for cosmological theories

Having attained a preliminary conclusion, it is now appropriate to return to the criteria for cosmological theories set out in §1.3. How does our approach relate to them?

The broad categories set forth there in Chapter 1 are: (1) satisfactory structure; (2) intrinsic explanatory power; (3) extrinsic explanatory power, or relatedness; (4) observational and experimental support. Our approach has focused on the fourth issue, namely observational support, in contrast to the approach of the astro-particle physicists, who particularly focus their approach on the third, namely the relation to a set of particle physics ideas and theories. Their approach is primarily model-based; ours is primarily data-based (cf. Matravers *et al.* 1995). Both approaches embody the first and second criteria to some degree, their approach perhaps more than ours.

We acknowledge the attraction and power of linking in a theory of the universe to other areas of physics. Indeed our approach has utilised knowledge from many areas of physics and astrophysics, which is tightly integrated in to our analysis. Where the major difference arises is that we have not placed a lot of weight on theories from the particle physics area that represent major extrapolations from well-established physics, and which we regard as weakly established experimentally. Indeed there is a major problem, from this viewpoint, with much of modern theoretical physics: a vast and elaborate structure has been created that has not yet attained experimental support, and indeed in many cases cannot *ever* be tested against observation. It is important and useful to develop these theories, but their status should not be mistaken: they are working hypotheses that may or may not be true.

The major point, then, is that any model-based approach to cosmology inevitably has to rely on such ill-tested theories. This is the reason we regard it as important to develop an approach which places major emphasis on observations, independently of such theories. The results of this observationally based approach can then be used as a way of testing the soundness of the theories proposed. Thus in effect we see our analysis not as providing a final theory of cosmology, but as providing a best-possible data analysis against which any theoretical proposals must be tested.

Finally we reiterate a point made in Chapter 1: the need for open-mindedness in approaching cosmological issues (e.g. Stoeger 1987). A review of the historical record will show that only in very few cases have we succeeded in predicting phenomena before they were observed (Harwitt 1984). Furthermore we have often in the past resisted the implications that were being pointed to by important evidence; for example, this happened in the case of the expanding universe (Ellis 1990). Thus we refer again to the meta-principle that we should look at all options in an open way, considering the whole phase space of possibilities, before coming to final conclusions about cosmology. In our view this point is fundamental; and it provides a hedge against dogmatism and authoritarian attitudes that have led us astray many times in the past. We hope this book will in some small measure assist in maintaining such an open-minded attitude.

9.4 Future prospects

The case we have made for $\Omega_0 \simeq 0.2$ though strong is by no means watertight. All the requirements for $\Omega_0 = 1$ to be correct could, of course, actually turn out to be correct. Some of the evidence we took to support $\Omega_0 \simeq 0.2$ could likewise turn out to have been wrong, or at least poorly interpreted. So while we believe the balance lies in the direction we have argued, the case is not proven beyond all reasonable doubt.

It is important, therefore, to consider those observations which may be crucial in proving the case finally, one way or the other. We have selected four, roughly in order of likely impact:

- The detection of a Doppler peak in the microwave background radiation.
- The experimental detection of a particle with the correct mass to contribute a large fraction of the critical density.
- A clear measurement of $q_0 = 0.5$ from Type Ia supernovae.
- Strong evidence from gravitational lensing of an error in cluster virial mass estimates.

The first of these is regarded by many cosmologists as the definitive result which will make them settle their wagers with each other. We would only like to add that it is possible that the temperature pattern measured by MAP or COBRAS/SAMBA looks nothing like any of the predictions from the favoured models, in which case estimation of cosmological parameters would be very difficult!

We have played down the role of direct dark matter detection experiments in this book as they seem to us to be long shots and, in any case, the extrapolation of the detection of objects in an earthly detector to the cosmological abundance of exotic particles is by no means solid (one would have to then consider the likely distribution of that particle in the universe). It would be compelling, however, if a well-known species (particularly the neutrino) turned out to have a finite mass. Most cosmologists would then accept the weak interaction freeze-out argument as essentially determining the contribution to Ω_0.

Perhaps the best prospect in the very near future is classical cosmology with Type Ia supernovae. The rate at which these are being detected suggests that errors on q_0 small enough to rule out flat Λ-dominated models will be acquired within a few years, before the CMB experiments fly. This is, of course, provided that no evolutionary or systematic effects come to light.

The cluster lensing arguments are very direct methods of probing the distribution of all gravitating matter in an object and are therefore less open to interpretation than many others. Lensing, particularly in the context of galaxy clusters, is a highly active field and one can hope for the many technical barriers to be overcome in the next few years. We placed this category last on our list, however, as this argument will still apply to relatively small scales and would not, for example, constrain the existence of a smooth component of the distribution.

9.5 The standard model of cosmology

We have now reached the point where we should rest our case, and wait for future observational developments to decide the verdict: we have stressed all along that this issue must, in the end, be settled empirically if it is to be settled at all.

We do, however, have one further point to make, motivated by some misleading statements in the press about there being a crisis in modern cosmology and that the 'standard model' of cosmology is wrong. Standard models are extremely important to a field of science, in that they encourage the construction of critical tests within the framework of the model to check its consistency with observations: when such consistency is lacking, one has to either change or discard the standard model. The central issue we must face here is what should be regarded as the 'standard model' of cosmology. What does it include? Clearly there is agreement that it includes

(a) expansion of the universe,
(b) from a hot big bang,
(c) where nucleosynthesis of the light elements took place,
(d) resulting in the CMB as relic radiation.

Some workers would like to include as part of the standard model additionally the assumptions of

(e) inflation in the past,
(f) leading to a critical density today, and
(g) with primeval quantum fluctuations providing the seeds from which galaxy formation took place...
(h) ...by gravitational instability of cold dark matter (or perhaps a mixture of hot and cold dark matter).

We argue *firstly* that while statements (a)–(d) are essential components of the standard model, the latter properties (e)–(h), exciting as they are, should not be regarded as such, but rather as major contenders to become part of the standard model in the future when they have achieved convincing experimental confirmation. This suggestion would save the cosmological community from some of the embarrassment caused by misleading articles recently suggesting that COBE and/or recent large-scale structure data has undermined our standard understanding of cosmology, when all it has done is shown that the simplest models satisfying (e)–(h) are unviable.

Secondly, we argue that whether elements (e)–(h) are regarded as part of the standard model or not, they should be tested by all possible observations in an open-minded way, just as the more basic elements (a)–(d) should.

In particular, we stress that Ω_0 is essentially a parameter describing the behaviour of the standard cosmological models, i.e. those incorporating (a)–(d). Being a parameter, it should be determined by fitting to the data, together with other parameters such as the Hubble parameter H_0 and those belonging to (e)–(h): the bias parameter b, and the primordial spectral index n. As we have shown, there is enough leeway in the present observations to allow a low-density model. Those cosmologists who take it for granted that we live in a high-density universe – and there have been many – may turn out to be profoundly mistaken. Indeed, we suggest that on evaluating all the data, $\Omega_0 \simeq 0.2$–0.3 is indicated – and that this should be adopted by cosmologists as the observationally preferred range of values, until the evidence indicates otherwise.

References

Alcock, C. *et al.* 1993, Nature, **365**, 621.
Alexander, T. 1995, MNRAS, **274**, 909.
Alpher, R.A., Bethe, H.A. and Gamow, G. 1948, Phys. Rev., **73**, 803.
Alpher, R.A. and Herman, R.C. 1948, Nature, **162**, 774.
Applegate, J.H. and Hogan, C.J. 1985, Phys. Rev., **D31**, 3037.
Arp, H.C., Burbidge, G., Hoyle, F., Narlikar J.V. and Wickramasinghe, N.C. 1990, Nature, **346**, 807.
Auborg, P. *et al.* 1993, Nature, **365**, 623.
Baade, W. 1956, PASP, **68**, 5.
Babul, A. and White, S.D.M. 1991, MNRAS, **253**, 31P.
Bahcall, J.N. 1984, ApJ, **276**, 169.
Bahcall, N.A. 1977, Ann. Rev. Astr. Astrophys., **15**, 505.
Bahcall, N.A. and Cen, R. 1992, ApJ, **398**, L81.
Bahcall, N.A., Lubin, L.M. and Doorman, V. 1995, ApJ, **447**, L81.
Bahcall, N.A. and Soneira, R.M. 1983, ApJ, **270**, 20.
Bahcall, N.A. and Tremaine, S. 1988, ApJ, **326**, L1.
Bardeen, J.M. 1980, Phys. Rev. **D22**, 1882.
Bardeen, J. M., Bond, J.R., Kaiser, N. and Szalay, A.S. 1986, ApJ, **304**, 15.
Barrow, J.D. 1976, MNRAS, **175**, 359.
Barrow, J.D. and Tipler, F.J. 1986, *The Anthropic Cosmological Principle*, Oxford University Press, Oxford.
Bartelmann, M., Ehlers, J. and Schneider, P. 1993 Astr. Astrophys., **280**, 351.
Bartlett, J.G., Blanchard, A., Silk, J., and Turner, M.S. 1995, Science **267**, 980.
Belinsky, A., Grishchuk, L.P., Khalatnikov, I.M. and Zel'dovich, Ya. B. 1985, Phys. Lett. **155B**, 232.
Bennett, C.L. *et al.* 1996, ApJ, **464**, L1.
Bernstein, J. 1988, *Kinetic Theory in the Expanding Universe*, Cambridge University Press, Cambridge.
Bernstein, J., Brown, L.S. and Feinberg, G. 1988, Rev. Mod. Phys., **61**, 25.

Bertotti, B. 1966, Proc. R. Soc., **294**, 195.
Bertschinger, E. and Dekel, A. 1989, ApJ, **336**, L5.
Bertschinger, E., Dekel, A., Faber, S.M. and Burstein, D. 1990, ApJ, **364**, 370.
Bertschinger, E. and Gelb, J.M. 1991, Computers in Physics, **5**, 164.
Bildhauer, S. and Futamase, T. 1991, Gen. Rel. Grav., **23**, 1251.
Binney, J. 1993. Nature, **366**, 406.
Biviano, A., Girardi, M., Giuricin G., Mardirossian, F. and Mezzetti, M., 1993, ApJ, **411**, L13.
Blandford, R.D. and Kochnanek, C. 1987, in *Dark Matter in the Universe*, eds. Bahcall, J.N., Piran, T. and Weinberg, S., World Scientific, Singapore, p. 133.
Boesgaard, A.M. and Steigman, G. 1985, Ann. Rev. Astr. Astrophys., **23**, 319.
Boldt, E. 1987, Phys. Rep., **146**, 215.
Bolte, M. and Hogan, C.J. 1995, Nature, **376**, 399
Bond, J.R., Carr, B.J. and Arnett, W.D. 1983, Nature, **304**, 514.
Bond, J.R. and Efstathiou, G. 1984, ApJ, **285**, L45
Bond, J.R. and Efstathiou, G. 1987, MNRAS, **226**, 655.
Bondi, H. 1960, *Cosmology*, Second Edition, Cambridge University Press, Cambridge.
Bonometto, S.A. and Pantano, O. 1993, Phys. Rep., **228**, 175.
Borgani, S., Lucchin, F., Matarrese, S. and Moscardini L., 1996, MNRAS, **280**, 749.
Bower, R.G., Coles, P., Frenk, C.S. and White, S.D.M. 1993, ApJ, **405**, 403.
Branch, D. and Tamman, G.A. 1992, Ann. Rev. Astron. Astrophys., **30**, 359.
Branchini E., Plionis M. and Sciama, D.W. 1996, ApJ, **461**, L17.
Brandenberger, R.H. 1985, Rev. Mod. Phys., **57**, 1.
Briel, U.G., Henry, J.P. and Böhringer, H. 1992, Astr. Astrophys., **259**, L31.
Broadhurst, T.J., Taylor, A.N. and Peacock, J.A. 1995, ApJ, **438**, 49.
Bruni, M., Dunsby, P.K. and Ellis, G.F.R. 1992, ApJ, **395**, 34.
Bucher, M., Goldhaber, A.S. and Turok, N. 1995, Phys. Rev. **D52**, 3314.
Buchert, T. and Ehlers, J. 1993, MNRAS, **264**, 375.
Burles, S. and Tytler, D. 1996, Science, submitted.
Burstein, D. 1990, Rep. Prog. Phys., **53**, 421.
Bunn, E.F., Scott, D. and White, M. 1995, ApJ, **441**, 9.
Byrne, J.P. *et al.* 1990, Phys. Rev. Lett., **65**, 289.
Caditz, D. and Petrosian, V. 1989, ApJ, **337**, L65.

Carlberg, R.G., Yee, H.K.C. and Ellingson, E. 1994, ApJ, **437**, 63.
Carlberg, R.G., Yee, H.K.C., Ellingson, E., Abraham, R., Gravel, P., Morris S. and Pritchet, C.J. 1996, ApJ, **462**, 32.
Carr, B.J. 1994, Ann. Rev. Astr. Astrophys., **32**, 531.
Carswell, R.F., Rauch, M., Weymann, R.J., Cooke, A.J. and Webb, J.K. 1994, MNRAS, **268**, L1.
Carswell, R.F. *et al.* 1996, MNRAS, **278**, 506.
Cayón, L., Martínez-Gonzalez, E., Sanz, J.L., Sugiyama, N. and Torres, S. 1996, MNRAS, **279**, 1095.
Cen, R. 1992, ApJS, **78**, 341.
Chaboyer, B., Demarque, P., Kernan, P.J. and Krauss, L.M. 1996a, Science, **271**, 957.
Chaboyer, B., Demarque, P., Kernan, P.J., Krauss, L.M. and Sarajedini, A. 1996b, MNRAS, in press.
Charlton, J.C. and Turner, M.S. 1987, ApJ, **313**, 495.
Cho, H. T and Kantowski, R., 1994, Phys. Rev. **D50**, 6144.
Cole, S., Fisher, K.B. and Weinberg, D.H. 1995, MNRAS, **275**, 515.
Coleman, S. 1988, Nucl. Phys., **B310**, 643.
Coles, P. 1992, Comments on Astrophysics, **16**, 45.
Coles, P. 1993, MNRAS, **262**, 1065.
Coles, P. and Ellis, G.F.R. 1994, Nature, **370**, 609.
Coles, P. and Lucchin, F. 1995, *Cosmology: The Origin and Evolution of Cosmic Structure*, John Wiley & Sons, Chichester.
Collins, C.A., Nichol, R.C. and Lumsden, S.L. 1992, MNRAS, **254**, 295.
Copi, C.J., Schramm, D.N. and Turner, M.S. 1995a, Science, **267**, 192.
Copi, C.J., Schramm, D.N. and Turner, M.S. 1995b, Phys. Rev. Lett., **75**, 3981.
Coule, D.H. 1995, Class. Quant. Grav., **12**, 455.
Cousins, R.D. 1995, Am. J. Phys., **63**, 398.
Cowie, L.L. 1987, in *The Post-Recombination Universe*, eds. Kaiser, N. and Lasenby, A.N, Kluwer, Dordrecht, p. 1.
Cowsik, R. and McClelland, J. 1972, Phys. Rev. Lett., **29**, 669.
Cox, R.T. 1946, Am. J. Phys., **14**, 1.
Dabrowski, Y., Lasenby, A. and Saunders, R., 1995, MNRAS, **277**, 753.
Dalcanton, J.J., Canizares, C.R., Granados, A., Steidel, C.C. and Stocke, J.R., 1994, ApJ, **424**, 550.
Davis, M., Efstathiou, G., Frenk, C.S. and White, S.D.M. 1985, ApJ, **292**, 371.
Davis, M. and Peebles, P.J.E. 1983, ApJ, **267**, 465.
Dekel, A. 1986, Comments on Astrophysics, **11**, 235.
Dekel, A. 1994, Ann. Rev. Astr. Astrophys., **32**, 371.

Dekel, A., Bertschinger, E. and Faber, S.M. 1990, ApJ, **364**, 349.
Dekel, A., Bertschinger, E., Yahil, A., Strauss, M.A., Davis, M. and Huchra, J.P. 1993, ApJ, **412**, 1.
Dekel, A. and Rees, M.J. 1987, Nature, **326**, 455.
de Lapparent, V., Geller, M.J. and Huchra, J.P. 1986, ApJ, **302**, L1.
de Oliveira Costa, A. and Smoot, G.F. 1995, ApJ, **448**, 477.
de Sitter, W. 1917, MNRAS, **78**, 3.
de Vaucouleurs, G. 1961, Science, **157**, 203.
Dicke, R.H. and Peebles, P.J.E. 1979, in *General Relativity. An Einstein Centenary Survey*, eds. Hawking, S.W. and Israel, W., Cambridge University Press, Cambridge, p. 504.
Dicke, R.H., Peebles P.J.E., Roll, P.G. and Wilkinson, D.T. 1965, ApJ, **142**, 414.
Dimopolous S., Esmailzadeh, R., Hall, L.J. and Starkman, G.D. 1988a, Phys. Rev. Lett., **60**, 7.
Dimopolous S., Esmailzadeh, R., Hall, L.J. and Starkman, G.D. 1988b, ApJ, **330**, 545.
Dunlop, J., Peacock, J.A., Spinrad, H., Rey, A., Jimenez, R., Stern, D. and Windhorst, R. 1996, Nature, **381**, 581.
Dydak, F. 1991, in *Proceedings of 25th International Conference on High Energy Physics*, ed. Rhua, K.K., World Scientific, Singapore.
Dyer, C.C. 1976, MNRAS, **175**, 429.
Efstathiou, G. 1990, in *Physics of the Early Universe, Proceedings of 36th Scottish Universities Summer School in Physics*, eds. Peacock, J.A., Heavens, A.F. and Davies, A.T., Edinburgh University Press, Edinburgh, p. 361.
Efstathiou, G. 1995, MNRAS, **274**, L73.
Efstathiou, G. and Bond, J.R. 1986, MNRAS, **218**, 103.
Efstathiou, G. and Bond, J.R. 1987, MNRAS, **227**, 38P.
Efstathiou, G., Ellis, R.S. and Peterson, B.A. 1988, MNRAS, **232**, 431.
Efstathiou, G., Sutherland, W.J. and Maddox, S.J. 1990, Nature, **348**, 705.
Efstathiou, G. *et al.* 1990, MNRAS, **247**, 10P.
Einstein, A. 1917, Sitzungsberichte der Preussischen. Akad. d. Wissenschaften, 142.
Einstein, A., 1955, *The Meaning of Relativity*, Fifth Edition, Princeton University Press, Princeton.
Eke, V.R., Cole, S. and Frenk, C.S. 1996, MNRAS, **282**, 263.
Ellis, G.F.R. 1975, QJRAS, **16**, 245.
Ellis, G.F.R. 1984a. Ann. Rev. Astr. Astrophys., **22**, 157.
Ellis, G.F.R., 1984b, in *General Relativity and Gravitation*, eds. Bertotti, B. *et al.*, Reidel, Dordrecht, p. 215.

Ellis, G.F.R., 1988, Class. Qu. Grav., **5**, 891.
Ellis, G.F.R. 1990, in *Modern Cosmology in Retrospect*, eds. Bertotti, B., Balbinto, R., Bergia, S. and Messina., A., Cambridge University Press, Cambridge, p. 97.
Ellis, G.F.R. 1991, in *Gravitation, Proceedings of Banff Summer Research Institute on Gravitation*, eds. Mann, R. and Wesson, P., World Scientific, Singapore, p. 3.
Ellis, G.F.R. 1993, QJRAS, **34**, 315.
Ellis, G.F.R. and Baldwin, J. 1984, MNRAS, **206**, 377.
Ellis, G.F.R. and Brundrit, G.B. 1979, QJRAS, **20**, 37.
Ellis, G.F.R. and Bruni, M. 1989, Phys. Rev. **D40**, 1804.
Ellis, G.F.R., Ehlers, J., Börner, G., Buchert, T., Hogan, C.J., Kirshner R.P., Press, W.H., Raffelt, G., Thieleman, F.-K., van den Bergh, S. 1996, in Dahlem Konferenzen Report on *The Evolution of the Universe*, John Wiley & Sons, New York, in press.
Ellis, G.F.R. and Jaklitsch, M.J., 1989, ApJ, **346**, 601.
Ellis, G.F.R., Lyth, D.H. and Mijic, M.B. 1991, Phys. Lett., **B271**, 52.
Ellis, G.F.R. and Madsen, M.S. 1991, Class. Qu. Grav., **8**, 667.
Ellis, G.F.R. and Schreiber, G. 1986, Phys. Rev. Lett., **A115**, 97.
Ellis, G.F.R. and Stoeger, W.R. 1987, Class. Qu. Grav., **4**, 1679.
Ellis, G.F.R. and Tavakol, R. 1994, Class. Qu. Grav., **11**, 675.
Ellis, G.F.R. and Tivon, G., 1985, Observatory, **105**, 189.
Ellis, G.F.R. and Williams, R.M., 1988. *Flat and Curved Space-times*, Oxford University Press, Oxford.
Ellis, R.S. 1993, Ann. NY. Acad. Sci., **688**, 207.
Evrard, G. and Coles, P. 1995, Class. Qu. Grav. **12**, L93.
Faber, S.M. and Gallagher, J.S. 1979, Ann. Rev. Astr. Astrophys., **17**, 135.
Fahlman, G., Kaiser, N., Squires G. and Woods, D. 1994, ApJ, **437**, 56
Fisher, K.B., Davis, M., Strauss, M.A., Yahil, A. and Huchra, J.P. 1994, MNRAS, **267**, 427.
Fisher, K.B., Sharf, C.A. and Lahav, O. 1994, MNRAS, **266**, 219.
Fort, B. and Mellier, Y. 1994, Astr. and Astrophys. Rev., **5**, 239.
Freedman, W.L. *et al.* 1994, Nature, **371**, 757.
Fukugita, M., Hogan, C.J. and Peebles, P.J.E. 1993, Nature, **366**, 309.
Fukugita, M. and Turner, E.L. 1991, MNRAS, **253**, 99.
Futamase, T. 1989, MNRAS, **237**, 187.
Gamow, G. 1946, Phys. Rev. **70**, 572.
Garrett, A.J.M. and Coles, P. 1993, Comments on Astrophysics, **17**, 23.
Gibbons, G. W., Hawking, S. W. and Stewart, J.M. 1987, Nucl. Phys., **B281**, 736.
Gott, J.R. 1982, Nature, **295**, 304.

Gott, J.R., Gunn, J.E., Schramm, D.N. and Tinsley, B.M. 1974, ApJ, **194**, 543.
Gottlöber, S., Mücket, J.P., Starobinsky, A.A. 1994, ApJ, **434**, 417.
Gouda, N., Sugiyama, N. and Sasaki, M. 1991a, ApJ, **372**, L49.
Gouda, N., Sugiyama, N. and Sasaki, M. 1991b, Prog. Theor. Phys., **85**, 1023.
Gunn, J.E. and Peterson, B.A. 1965, ApJ, **142**, 1633
Gurvits, L.I. 1994, ApJ, **425**, 442.
Guth, A.H., 1981, Phys. Rev., **D23**, 347.
Guzman, R. and Lucey, J.R. 1993, MNRAS, **263**, L47.
Hamilton, A.J.S. 1992, ApJ, **385**, L5.
Hamilton, A.J.S. 1993, ApJ, **406**, L47.
Hamuy, M., Philips, M.M., Maza, J. Suntzeff, N.B., Schommer, R.A. and Aviles, R. 1995, Astr. J., **109**, 1.
Hancock, S. *et al.* 1993, Nature, **367**, 333.
Harrison, E.R. 1970, Phys. Rev., **D1**, 2726.
Harwitt, M. 1984, *Cosmic Discovery*, MIT Press, Cambridge, Mass.
Hawking, S.W. and Page, D.N. 1988, Nucl. Phys., **B298**, 789.
Hawkins, M.R.S. 1993, Nature, **366**, 242.
Heavens, A.F. and Taylor A.N. 1995, MNRAS, **275**, 483.
Hendry, M.A. and Tayler, R.J. 1996, Contemporary Physics, **37**, 263.
Hogan, C.J., Kaiser, N. and Rees, M.J. 1982, Phil. Trans. R. Soc., **A307**, 97.
Holtzman, J. 1989, ApJS, **71**, 1.
Hoyle, F. 1959, in *Paris Symposium on Radio Astronomy*, IAU No. 9, ed. Bracewell, R., p. 529.
Hoyle, F. and Tayler, R.J. 1964, Nature, **203**, 1108.
Hu, W. and Sugiyama, N. 1995, Phys. Rev. **D51**, 2599.
Hu, W., Sugiyama, N. and Silk, J. 1996, Nature, in press.
Hubble, E. 1926, ApJ, **64**, 321.
Hubble, E. 1929, Proc. Nat. Acad. Sci., **15**, 168
Hubble, E. 1934, ApJ, **79**, 8.
Hubble, E. 1936, ApJ, **84**, 517.
Hubner, P. and Ehlers, J. 1991, Class. Qu. Grav., **8**, 333.
Hudson, M.J. 1993, MNRAS, **265**, 72.
Islam, J.N. 1983, *The Ultimate Fate of the Universe*, Cambridge University Press, Cambridge.
Izatov, Y.I., Thuan, T.-X., Lipovetsky, V.A. 1994, ApJ, **435**, 647.
Janis, A. 1986, Am. J. Phys., **54**, 1008.
Jaynes, E.T. 1969, IEEE Trans. System Science Cybern., **SSC-4**, p. 227
Jeans, J.H. 1902, Phil. Trans. R. Soc., **199A**, 1.

Jeffreys, H. 1939, *Theory of Probability*, Clarendon Press, Oxford.
Jimenez, R., Thejll, P., Jorgensen, U.G., MacDonald, J. and Pagel, B.E.J. 1996, MNRAS, 282, 926.
Jones, M. *et al.* 1993, Nature, **365**, 320.
Kaiser, N. 1984, ApJ, **284**, L9.
Kaiser, N. 1987, MNRAS, **227**, 1.
Kaiser, N., Efstathiou, G., Ellis, R.S., Frenk, C.S., Lawrence, A., Rowan-Robinson, M. and Saunders, W. 1991, MNRAS, **252**, 1.
Kaiser, N. and Lahav, O. 1989, MNRAS, **237**, 129.
Kaiser, N. and Silk, J. 1987, Nature, **324**, 529.
Kaiser, N. and Squires, G. 1993, ApJ, **404**, 441.
Kamionkowski, M. and Spergel, D.N. 1994, ApJ, **432**, 7.
Kamionkowski, M., Spergel, D.N. and Sugiyama N. 1994, ApJ, **426**, L57.
Katz, N., Weinberg, D.H., Hernquist, L. and Miralda-Escudé, J., 1996, ApJ, **457**, L57.
Kauffman, G. and Charlot, S. 1994, ApJ, **430**, L97.
Kellerman, K.I. 1993, Nature, **361**, 134.
Klypin, A.A., Borgani, S., Holtzman, J. and Primack, J.R. 1995, ApJ, **444**, 1.
Klypin, A.A., Holtzmann, J.A., Primack, J. and Regös, E. 1993, ApJ, **416**, 1.
Kneib, J.-P., Mellier, Y., Fort, B. and Mathez, G. 1993, Astr. Astrophys., **273**, 367.
Kochanek, C.S. 1993, ApJ, **419**, 12.
Kolb, E.W. and Turner, M.S. 1990, *The Early Universe*, Addison-Wesley, Redwood City.
Kolokotronis V., Plionis M., Coles P., Borgani S. and Moscardini L. 1996, MNRAS, **280**, 186.
Koo, D.C. 1988, in *The Epoch of Galaxy Formation*, eds. Frenk, C.S., Ellis, R.S., Shanks, T., Heavens, A.F. and Peacock, J.A., Kluwer, Dordrecht, p. 71.
Krauss, L.M. and Turner, M.S. 1995, Gen. Rel. Grav., **27**, 1137.
Lacey, C.G. and Cole, S. 1993, MNRAS, **262**, 627.
Lahav, O. 1987, MNRAS, **225**, 213.
Lahav, O., Kaiser, N. and Hoffman, Y. 1990, ApJ, **352**, 448.
Lahav, O., Rowan-Robinson, M. and Lynden-Bell, D. 1988, MNRAS, **234**, 677.
Lanzetta, K.M., Wolfe, A.M. and Turnshek, D.A. 1995, ApJ, **440**, 435.
Lanzetta, K.M., Yahil, A. and Fernandez-Soto, A. 1996, Nature, **381**, 759.
Lauer, T. and Postman, M. 1994, ApJ, **425**, 418.

Lee, B.W. and Weinberg, S. 1977. Phys. Rev. Lett., **39**, 165.
Liddle, A.R. and Lyth, D.H. 1993, Phys. Rep., **231**, 1.
Liddle, A.R., Lyth, D.H., Roberts, D. and Viana, P.T.P. 1996, MNRAS, **278**, 644.
Lidsey, J.E. and Coles, P. 1992, MNRAS, **258**, 57P.
Linde, A.D. 1982, Phys. Lett., **B116**, 335.
Linde, A. 1990, *Inflation and Quantum Cosmology*, Academic Press, Boston.
Linde, A. and Mezhlumian, A. 1995, Phys. Rev., **D52**, 4267.
Linsky, J.L. *et al.* 1993, ApJ, **402**, 694.
Linsky, J.L. *et al.* 1995, ApJ, **451**, 335.
Loeb, A. and Mao, S. 1994, ApJ, **435**, L109.
Loh, E.D. and Spillar, E.J. 1986, ApJ, **307**, L1.
Lubin, P.M., Epstein, G.L. and Smoot, G.F. 1983, Phys. Rev. Lett., **50**, 616.
Lucchin, F. and Matarrese, S. 1985, Phys. Lett., **164B**, 282.
Lynden-Bell, D., Lahav, O. and Burstein, D. 1989, MNRAS, **241**, 325.
Lynden-Bell, D. and Wood, R. 1968, MNRAS, **138**, 495.
Ma, C.P. and Bertschinger, E. 1994, ApJ, **434**, L5.
Maddox, S.J., Efstathiou, G., Sutherland, W.J. and Loveday, J. 1990, MNRAS, **242**, 43P.
Madsen, M. and Ellis, G.F.R. 1988, MNRAS, **234**, 67.
Malaney, P. and Mathews, G. 1993, Phys. Rep., **229**, 147.
Mao, S. and Kochanek, C.S. 1994, MNRAS, **268**, 569.
Maoz, D. and Rix, H.-W. 1993, ApJ, **416**, 425.
Mather, J.C. *et al.* 1994, ApJ, **420**, 439.
Matravers, D.R., Ellis, G.F.R. and Stoeger, W.R. 1995, QJRAS, **36**, 29.
Metcalfe, N., Shanks, T., Campos, A., Fong, R. and Gardner, J.P. 1996, Nature, **383**, 236.
Milgrom, M. 1983, ApJ, **270**, 365.
Milne, E.A. 1936, Proc. Roy. Roc. Lond. **A154**, 22.
Mohr, J.J., Evrard, A.E., Fabricant, D.G. and Geller, M.J. 1995, ApJ, **447**, 8.
Mo, H.J. and Miralda-Escudé, J. 1994, ApJ, **430**, L25.
Molaro, P., Primas, F. and Bonifacio, P. 1995, Astr. Astrophys., **295**, L47.
Moscardini, L., Branchini, E., Tini Brunozzi, P., Borgani, S., Plionis, M. and Coles, P. 1996, MNRAS, **282**, 384.
Narlikar, J.V. 1993, *Introduction to Cosmology*, Second Edition, Cambridge University Press, Cambridge.
Newsam, A., Simmons, J.F.L. and Hendry, M.A. 1995, Astr. Astrophys., **294**, 627.

Nugent, P., Philips, M., Baron, E., Branch D. and Hauschildt, P. 1995, ApJ, **455**, L147.
Olive, K.A. 1990, Phys. Rep., **190**, 307.
Olive, K.A. and Schramm, D.N., 1993, Nature, **360**, 439.
Olive, K.A. and Scully, S.T. 1995, Int. J. Mod. Phys. **A11**, 409.
Olive, K.A. and Steigman, G. 1995, ApJS, **97**, 49.
Oort, J.H. 1932, Bull. Astron. Inst. Neth., **6**, 249.
Ostriker, J.P. and Steinhardt, P.J. 1995, Nature, **377**, 600.
Ostriker, J.P. and Vishniac, E.T. 1986, ApJ, **306**, L5.
Overbye, D. 1993, *Lonely Hearts of the Cosmos*, Picador, London.
Paczynski, B. 1986a, ApJ, **304**, 1.
Paczynski, B. 1986b, ApJ, **308**, L43.
Padmanabhan, T. 1990, Phys. Rep., **188**, 285.
Pagel, B.E.J., Simonson, E.A., Terlevich, R.J. and Edmunds, M.G. 1992, MNRAS, **255**, 325.
Panek, M. 1986, Phys. Rev., **D34**, 416.
Partridge, R.B. 1995, *3K: The Cosmic Microwave Background Radiation*, Cambridge University Press, Cambridge.
Peacock, J.A. 1991, MNRAS, **253**, 1P.
Peacock, J.A. 1996, MNRAS, in press.
Peacock, J.A. and Dodds, S.J. 1994, MNRAS, **267**, 1020.
Peebles, P.J.E. 1965, ApJ, **146**, 542.
Peebles, P.J.E. 1971, *Physical Cosmology*, Princeton University Press, Princeton.
Peebles, P.J.E. 1980, *The Large-scale Structure of the Universe*, Princeton University Press, Princeton.
Peebles, P.J.E. 1984, Science, **224**, 1385.
Peebles, P.J.E. 1986, Nature, **321**, 27.
Peebles, P.J.E. 1987, ApJ, **315**, L73.
Peebles, P.J.E. 1993, *Principles of Physical Cosmology*, Princeton University Press, Princeton.
Peebles, P.J.E., Schramm, D.N., Turner, E.L. and Kron, R.G. 1991, Nature, **353**, 769.
Peebles, P.J.E. and Yu, J.T. 1970, ApJ, **162**, 815.
Penzias, A.A. and Wilson, R.W. 1965, ApJ, **142**, 419.
Perlmutter, S. *et al.* 1995, ApJ, **440**, L41.
Perlmuttter, S. *et al.* 1996, ApJ, submitted.
Persic, M. and Salucci, P. 1992, MNRAS, **258**, 14P.
Pierce, M.J., Welch, D.L., McClure R.D., van den Bergh, S., Racine R. and Stetson P.B. 1994, Nature, **371**, 385.
Pierpaoli, E., Coles, P., Bonometto, S., and Borgani, S. 1996, ApJ, **470**, 92.

Plionis M., Coles, P. and Catelan, P. 1992, MNRAS, **262**, 465.
Plionis, M. and Valdarnini, R. 1991, MNRAS, **249**, 46.
Press, W.H. and Schechter, P.L. 1974, ApJ, **187**, 425.
Ratra, B., Banday, A.J., Gorski, K.M. and Sugiyama, N. 1996, preprint.
Ratra, B. and Peebles, P.J.E. 1995, Phys. Rev. **D52**, 1837.
Rees, M.J. 1985, MNRAS, **213**, 75P.
Rees, M.J. 1995. *Perspectives in Astrophysical Cosmology*, Cambridge University Press, Cambridge.
Rees, M.J. and Sciama, D.W. 1968, Nature, **217**, 511.
Richstone, D., Loeb, A. and Turner, E.L. 1992, ApJ, **393**, 477.
Riess, A.G., Press, W.H. and Kirshner, R. P., 1995, ApJ, **438**, L17.
Rothman, T. and Ellis, G.F.R. 1987, Observatory, **107**, 24.
Rowan-Robinson, M. 1985, *The Cosmological Distance Ladder*, W.H. Freeman, New York.
Rowan-Robinson, M. *et al.* 1990, MNRAS, **247**, 1.
Rubin, V.C., Ford, W.K. and Thonnard, N. 1980, ApJ, **238**, 471.
Rugers, M. and Hogan, C.J., 1996a, ApJ, 459, L1.
Rugers, M. and Hogan, C.J., 1996b, Astr. J., **111**, 2135.
Sachs, R.K. and Wolfe, A.M. 1967, ApJ, **147**, 73.
Sahni, V. and Coles, P. 1995, Phys. Rep., **262**, 1.
Sandage, A.R. 1961, ApJ, **133**, 355.
Sandage, A.R. 1968, Observatory, **88**, 91.
Sandage, A.R. 1970, Physics Today, **23**, 34.
Sandage, A.R. 1982, ApJ, **252**, 553.
Sandage, A.R. 1988, Ann. Rev. Astr. Astrophys., **26**, 561.
Sandage, A.R. 1995, in *The Deep Universe*, Saas-Fee Advanced Course 23, eds. Sandage, A.R., Kron, R.G. and Longair, M.S., Springer-Verlag, Berlin.
Sandage, A. and Tamman, G.A. 1990, ApJ, **365**, 1.
Sarazin, C. 1986, Rev. Mod. Phys., **58**, 1.
Saunders, W. *et al.* 1991, Nature, **349**, 32.
Scaramella, R., Vettolani, G. and Zamorani, G. 1991, ApJ, **376**, L1.
Sciama, D. W. 1990, ApJ, **364**, 549.
Sciama, D.W. 1993, *Modern Cosmology and the Dark Matter Problem*, Cambridge University Press, Cambrige.
Schmidt, M. 1968, ApJ, **151**, 393.
Schneider, P., Ehlers, J, and Falco, E.E. 1992, *Gravitational Lenses*, Springer-Verlag, Berlin.
Scott, D., Silk, J. and White, M. 1995, Science, **268**, 829.
Scott, P.F. and Ryle, M. 1961, MNRAS, **122**, 389.

Scully, S.T., Cassé, M., Olive, K.A., Schramm, D.N., Truran, J.W. and Vangioni-Flam, E. 1996, ApJ, **464**, 960.
Serna, A. and Alimi, J.-M. 1996a, Phys. Rev. **D53**, 3074.
Serna, A. and Alimi, J.-M. 1996b, Phys. Rev. **D53**, 3087.
Shapley, H. 1934, MNRAS, **94**, 791.
Shectman, S.A., Landy, S.D., Oemler, A., Tucker, D.L., Lin, H., Kirshner, R.P. and Schechter, P.L. 1996, ApJ, in press.
Shi, X., Widrow, L.M. and Dursi, L.J. 1996, MNRAS, **281**, 565.
Shirokov, M.F. and Fisher, I.Z. 1963, Soviet Astr., **6**, 699.
Sikivie, P. 1985, Phys. Rev. **D32**, 2988.
Skillman, E.D., Terlevich, R.J., Terlevich, E., Kennicutt, R.C. and Garnett, D.R. 1993, Ann. NY. Acad. Sci., **688**, 739.
Smail, I., Ellis, R.S. and Fitchett, M.J., 1995a, MNRAS, **270**, 245.
Smail, I., Ellis, R.S., Fitchett, M.J. and Edge, A.C. 1995b, MNRAS, **273**, 277.
Smail, I., Couch, W.J., Ellis, R.S. and Sharples, R.M. 1995c, ApJ, **440**, 501.
Smith, M.S., Kawano, L.H. and Malaney, R.A. 1993, ApJS, **85**, 219.
Smith, P.F. 1986, in *Cosmology, Astronomy and Fundamental Physics: Proceedings of 2nd ESO/CERN Symposium*, eds. Setti, G. and Van Hove, L., p. 237.
Smoot, G. F. *et al.* 1992, ApJ, **396**, L1.
Smoot, G.F., Tenorio, L., Banday, A.J., Kogut, A., Wright, E.L., Hinshaw, G. and Bennett, C.L. 1994, ApJ, **437**, 1.
Songaila, A., Cowie, L.L., Hogan, C.J. and Rugers, M. 1994, Nature, **368**, 599.
Spergel, D.N. 1996, in *Some Oustanding Questions in Astrophysics*, eds. Bahcall, J.N. and Ostriker, J.P., Princeton University Press, Princeton, in press.
Spergel, D.N. and Press, W.H. 1985, ApJ, **294**, 663.
Spite, M., Francois, P., Nissen, P.E. and Spite, F. 1996, Astr. Astrophys., **307**, 172.
Steigman, G., 1994, MNRAS, **269**, L53.
Steigman, G., Fields, B., Olive, K.A., Schramm, D.N. and Walker, T.P. 1993, ApJ, **415**, L35.
Steigman, G. and Tosi, M. 1995, ApJ, **453**, 173.
Stepanas, P.G. and Saha, P. 1995, MNRAS, **272**, L13
Stoeger, W.R. 1987, *Theory and Observational Limits in Cosmology*, Specola Vaticana, Citta del Vaticano.
Stoeger, W.R., Ellis, G.F.R. and Xu, C. 1994, Phys. Rev. **D49**, 1845.
Stoeger, W.R., Maartens, R. and Ellis, G.F.R. 1995a, ApJ, **443**, 1.

Stoeger, W.R., Xu, C.-M., Ellis, G.F.R. and Katz, M. 1995b, Phys. Rev., **D45**, 17.
Storrie-Lombardi, L.J., McMahon, R.G., Irwin, M.J. and Hazard, C. 1995, in *Proceedings of the ESO Workshop on QSO Absorption Lines*, in press.
Strauss, M. and Davis, M. 1998a, in Proceedings of 3rd IRAS Conference, *Comets to Cosmology*, ed. Lawrence, A., Springer-Verlag, Berlin, p. 361.
Strauss, M. and Davis, M. 1988b, in *Large-scale Motions in the Universe, Proceedings of Vatican Study Week*, eds. Rubin, V.C. and Coyne, G., Princeton University Press, Princeton, p. 256.
Strauss, M.A. and Willick, J.A. 1995, Phys. Rep., **261**, 271.
Sunyaev, R.A. and Zel'dovich, Ya.B. 1969, Comm. Astrophys. Space Phys., **4**, 173.
Sunyaev, R.A. and Zel'dovich, Ya.B. 1970, Astrophys. Sp. Sci., **7**, 3.
Suto, Y., Suginohara, T., Inagaki, Y. 1995, Prog. Theor. Phys., **93**, 839.
Symbalisky, E.M.D. and Schramm, D.N. 1981, Rep. Prog. Phys., **44**, 293.
Tanvir, N.R., Shanks, T., Ferguson, H.C. and Robinson, D.R.T. 1995, Nature, **377**, 27.
The, L. and White, S.D.M. 1986, Astr. J., **92**, 1248.
Thorne, K.S. 1967, ApJ, **148**, 51.
Tini Brunozzi P., Borgani S., Plionis M., Moscardini L. and Coles P. 1995, MNRAS, **277**, 1210.
Tolman, R.C. 1934, *Relativity, Thermodynamics and Cosmology*, Clarendon Press, Oxford. Reprinted, 1987, Dover, New York.
Traschen, J.E. 1984, Phys. Rev. **D29**, 1563.
Traschen, J.E. and Eardley, D.M. 1986, Phys. Rev., **D34**, 1665.
Trimble, V. 1987, Ann. Rev. Astr. Astrophys., **25**, 425.
Tsai, J.C. and Buote, D.A. 1996, MNRAS, 282, 77.
Turner, E.L. 1990, ApJ, **365**, L43.
Turner, E.L., Ostriker, J.P. and Gott, J.R., 1984, ApJ, **284**, 1.
Turner, M.S., Steigman, G. and Krauss, L.M. 1984, Phys. Rev. Lett., **52**, 2090.
Tyson, J.A., 1988, AJ, **96**, 1.
Tyson, J.A. and Seitzer, P. 1988, ApJ, **335**, 552.
Tyson, J.A., Valdes, F. and Wenk, R.A. 1990, ApJ, **349**, L1.
Tytler, D., Fan, X.-M. and Burles, S. 1996, Nature, **381**, 207.
Viana, P.T.P. and Liddle, A.R. 1996, MNRAS, **281**, 323.
Vittorio, N. and Silk, J. 1984, ApJ, **285**, L39.
Wainwright, J. and Ellis, G.F.R. 1996. *The Dynamical Systems Approach to Cosmology*, Cambridge University Press, Cambridge.

Walker, T.P., Steigman, G., Schramm, D.N., Olive, K.A. and Kang, K.S. 1991, ApJ, **376**, 51.
Walker, T.P., Steigman, G., Schramm, D.N., Olive, K.A. and Fields, B., 1993, ApJ, **413**, 562.
Walsh, D., Carswell, R.F. and Wymann, R.J. 1979, Nature, **279**, 381.
Wampler, E.J. *et al.* 1996, Astr. Astrophys., in press.
Weinberg, S. 1972, *Gravitation and Cosmology*, John Wiley & Sons, New York.
Weinberg, S. 1989, Rev. Mod. Phys., **6**, 1.
West, M.J., Jones, C. and Forman, W. 1995, ApJ, 451, L5.
White, M. and Bunn, E.F. 1995, ApJ, **450**, 477.
White, M., Scott, D. and Silk, J. 1994, Ann. Rev. Astr. Astrophys. **32**, 319.
White, S.D.M., Briel, U.G. and Henry, I.P. 1993, MNRAS, **261**, L8.
White, S.D.M., Efstathiou, G. and Frenk, C.S. 1993, MNRAS, **262**, 1023.
White, S.D.M., Navarro, J.F., Evrard, A.E. and Frenk, C.S. 1993, Nature, **366**, 429.
Wilson, G., Cole, S. and Frenk, C.S. 1996, MNRAS, **282**, 501.
Wilson, M.L. 1983, ApJ, **273**, 2.
Wolfe, A.M. 1993, Ann. NY. Acad. Sci., **688**, 281.
Wolfe A.M., Lanzetta K.M., Foltz, C.B. and Chaffee, F.H. 1995, ApJ, **454**, 698.
Wright, E. *et al.* 1992, ApJ, **396**, L7.
Wu, X.-P., Deng, Z., Zou, Z., Fang, Li-Z. and Qin, B. 1995, ApJ, **448**, L65.
Zaritsky, D., Smith, R., Frenk, C.S. and White, S.D.M. 1993, ApJ, **405**, 464.
Zel'dovich, Ya.B. 1972, MNRAS, **160**, 1P.
Zotov, N.V. and Stoeger, W.R. 1992, Class. Qu. Grav., **9**, 1023.
Zotov, N.V. and Stoeger, W.R. 1995, ApJ, **453**, 574.
Zwicky, F. 1933, Helvetica Physica Acta, **6**, 110.

Index

Printed in the United States
By Bookmasters